John Van Denburgh

The Reptiles of the Pacific Coast and Great Basin

An account of the species known to inhabit California, and Oregon, Washington,

Idaho and Nevada

John Van Denburgh

The Reptiles of the Pacific Coast and Great Basin
An account of the species known to inhabit California, and Oregon, Washington, Idaho and Nevada

ISBN/EAN: 9783337012014

Printed in Europe, USA, Canada, Australia, Japan

Cover: Foto ©berggeist007 / pixelio.de

More available books at **www.hansebooks.com**

THE REPTILES

PACIFIC COAST AND GREAT BASIN

AN ACCOUNT OF THE SPECIES KNOWN TO INHABIT

CALIFORNIA,

Oregon, Washington, Idaho and Nevada.

BY

JOHN VAN DENBURGH, PH. D.

Curator of the Department of Herpetology.

PREFACE.

The life histories of our reptiles remain almost entirely unknown. Herpetologists are so few, and reptiles so retiring, that, unless more general interest can be aroused, we may hope for very little light upon this important branch of the science. Many specialists in other departments of science, as well as hunters, farmers, students, and other intelligent men, however, are constantly in the field and might record observations of great interest were means of identification at hand. This paper has been prepared in the hope that it may stimulate those whose mode of life leads them into the woods and fields to study the ways of our reptiles—by no means the least interesting, if amongst the more humble, of the animals about us.

Although it has not been thought advisable to "popularize" the descriptions in the following pages, such characters as cannot be determined without dissection have been avoided. For this reason the synopses and characterizations of the higher groups are very artificial and are not intended to hold good if applied to extralimital species or genera. Likewise, superfamilies, subfamilies, and subgenera have not been introduced. While this is, thus, intended as a handbook for the more or less casual student, it is hoped that the professional herpetologist will find something of interest regarding the variation and distribution of our reptiles.

I wish especially to express my obligation to Dr. Leonhard Stejneger, who placed at my disposal the entire collection of reptiles belonging to the United

States National Museum; to Mr. Henry H. Hindshaw, Curator of the Museum of the University of Washington, who sent me a collection of species from western Washington; to Mr. John Fannin, Curator of the Provincial Museum, Victoria, B. C., who lent me the collection belonging to his institution; to Prof. W. E. Ritter, of the University of California, for a similar favor; to Dr. G. Baur; to Mr. Harold Heath; to Dr. W. W. Thoburn; to Mr. Douglas Van Denburgh; to Mr. Edward Hyatt; to Mr. J. M. Hyde; to Mr. J. O. Snyder; to Dr. G. Eisen; to many others; and most of all to Dr. Chas. H. Gilbert, who forwarded to me the unequaled collection of Californian reptiles belonging to Leland Stanford Junior University, and gathered for that institution by his untiring zeal.

THE AUTHOR.

San Francisco, April, 1897.

CONTENTS.

8   CONTENTS.

THE REPTILES OF THE PACIFIC COAST AND GREAT BASIN.

INTRODUCTION.

The term reptile is popularly applied to all cold-blooded vertebrates other than fishes. Thus used it includes two groups of animals which differ in many important respects. These are the batrachians and the reptiles proper; the former more closely allied to the fishes; the latter, to the birds.

The typical batrachians, such as most frogs, toads, salamanders, etc., lay their eggs in the water, and the young, for a time, breathe by means of gills, very much as do the fishes. Later on, they undergo a metamorphosis, during which the gills and other larval characteristics disappear, the tadpole assumes the form and structure of its parents and emerges from the water to breathe air and spend a greater or less portion of its life on land. The skin of batrachians* is not provided with scales, but is smooth or warty, very glandular, and often covered with a slimy secretion.

The true reptiles, such as alligators, turtles, lizards, and snakes, on the other hand, never lay their eggs in the water, even the marine species coming to land for this purpose. Their young never breathe by means of gills, but are hatched or born with the form and structure of the adult. The skin, except of some turtles, is covered with scales, and is dry, never slimy.

There are, also, many anatomical and embryological differences between the two classes, but these need not

---

*Except Cœcilians of tropical lands.

be mentioned here, since the batrachians will not be considered in the following pages. Our reptiles and batrachians may be distinguished by the following

SYNOPSIS OF CLASSES.

a.—Anal opening transverse or round; skin furnished with scales (varying from large plates to minute granules); or, if skin smooth (*Pelodiscus*), tail and claws present and jaws without teeth. (Turtles, lizards, snakes, etc.)..................................**Reptilia.**—p. 28.

a².—Anal opening longitudinal or round; skin smooth or warty, without scales; no claws.* (Frogs, toads, salamanders, newts, water-dogs, tadpoles, etc.)..**Batrachia.**

Long ago the reptiles were the rulers of the earth as the mammals are to-day. Huge monsters, most grotesquely fashioned, roamed over the land, while equally large reptiles swam in the seas, and in the air were great creatures whose bat-like wings, it is said, sometimes measured more than twenty feet from tip to tip. But of these monsters, none remain alive; only the smaller forms have survived. Living reptiles fall naturally into four groups or orders. One of these orders contains but a single lizard-like animal, the *Sphenodon* of New Zealand, interesting to the morphologist because of its generalized structure. The other three orders are numerously represented in the warmer portions of both the Old and New Worlds. They are: first, the alligators and crocodiles; second, the turtles; third, the lizards and the snakes.

The alligators and crocodiles are of chiefly tropical and subtropical distribution and do not enter the territory we are considering. The turtles are most numerous in moist regions, and, consequently, are represented on the Pacific Coast and in the Great Basin by few species. The lizards and snakes, on the contrary, find

---

* Tips of digits sometimes horny.

our warm, dry climate well adapted to their needs, and are very numerous. In the following pages I have admitted to the fauna of the states under consideration seventy-seven species and subspecies of reptiles belonging to thirty-seven genera, thirteen families, and two orders. Of these, three are turtles, forty are lizards, and thirty-four snakes.

While it is probable that no two of these species have exactly the same geographical limits, yet the ranges of certain species are, in a general way, conterminous with those not only of other reptiles but of other kinds of animals as well. Thus, if we map out the areas occupied by the different kinds of mammals, birds, reptiles, insects, plants, etc., we find that the boundaries of the ranges of many species are nearly coincident, so that in one area we have certain genera and species associated, while more or less closely related kinds inhabit adjoining districts. From such study of its animals and plants temperate North America has been divided into a number of life zones,* each of which may be subdivided into minor areas technically known as Faunæ.

When regarded from a herpetological standpoint, California may be divided into five minor life areas, each of which corresponds more or less closely with one of the chief physical areas of the State. Thus, one biologic area occupies the southeastern deserts, another the southern coast, a third the western slopes of the northern coastal ranges, a fourth a belt along the Sierra Nevada, and a fifth the great interior valleys of the Sacramento and San Joaquin together with their fringing foothills.

---

* On this subject see especially Allen, Bull. Am. Mus. N. H., IV, 1, 1892, pp. 199-244; Auk, X, 2, Apr., 1893; Merriam, N. A. Fauna, No, 3, 1890, and No. 5, 1891; Proc. Biol. Soc. Wash. VII, pp. 1-64, 1892; Nat. Geog. Mag., 1894; Rep. Sec. Agri., 1893, pp. 228, 229 (1894).

Each of these areas is characterized by the presence of certain species which do not live in the others, and the absence of other species peculiar to the adjoining Faunæ. Other species, though not restricted to one, conform more or less closely to the geographical limits of two or more of these life areas. Without stopping to speculate upon the causes which, severally or in combination, have operated to bring about this arrangement, let us consider each of these areas in detail, pointing out its faunal peculiarities and relations to the others.

The reptilian fauna of California, as at present known, is composed of seventy-one species and subspecies. The following table shows the area or areas which each of these is known to inhabit:

DISTRIBUTION OF CALIFORNIAN REPTILES.

SPECIES AND SUBSPECIES.	Sierra Nevada...	Northern Coast..	Valleys	Southern Coast.	Desert.........
Clemmys marmorata..................		x	x	x	
Chrysemys bellii (?)................			?		
Gopherus agassizii..................					x
Coleonyx variegatus					x
Dipsosaurus dorsalis					x
Uma notata					x
inornata.......................					x
Callisaurus ventralis..............					x
Crotaphytus baileyi................					x
wislizenii					x
silus.....................			x		
Sauromalus ater...................					x
Uta mearnsi				x	
stansburiana			x	x	x
graciosa.....................					x
symmetrica					x

DISTRIBUTION OF CALIFORNIAN REPTILES—Continued.

SPECIES AND SUBSPECIES.	Sierra Nevada.	Northern Coast.	Valleys.	Southern Coast.	Desert.
Sceloporus graciosus	x			*	
occidentalis		x		x	
biseriatus			x	x	?
magister					x
orcutti				x	
Phrynosoma douglassii			?		
blainvillii				x	
frontale			x		
platyrhinos					x
m'callii					x
Gerrhonotus scincicauda		x	x	x	
burnettii		x			
principis (?)	?				
palmeri	x				
Anniella pulchra			x	x	
nigra		x			
Xantusia vigilis					x
henshawi				x	
riversiana				?	
Cnemidophorus tigris					x
undulatus				x	
stejnegeri				x	
Verticaria hyperythra beldingi				x	
Eumeces skiltonianus	x	x	x	x	
gilberti	x				
Siagonodon humilis					x
Lichanura roseofusca				x	
Charina bottæ	x	x	x		x
Chilomeniscus ephippicus					x
Chionactis occipitalis					x
Contia mitis		x	x		
Diadophis amabilis	x	x	x	x	
Lampropeltis zonata	x	x	?	?	
boylii	?		?	x	x
californiæ			x	x	
Rhinocheilus lecoutei			x	x	x
Tantilla eiseni				x	
Hypsiglena ochrorhynchus				x	x
Salvadora grahamiæ			x	x	
Bascanion constrictor vetustum	x	x	x	x	
flagellum frenatum			x	x	x
laterale			x	x	
tæniatum			?		x

* In mountains only.

DISTRIBUTION OF CALIFORNIAN REPTILES—Concluded.

Species and Subspecies.	Sierra Nevada.	Northern Coast.	Valleys.	Southern Coast.	Desert.
Arizona elegans.............................				x ..	.. ? ..
Pituophis catenifer.....		x .	x .	x .	
deserticola.............					.. x ..
Thamnophis parietalis	x ..	x .	x .	.. x .	
elegans..................	.. x ..	x .	x .		.. ..
vagrans	.. x .		x .		x ..
hammondii		x .	x ..	x ..	
Crotalus lucifer.	x ..	x .	x .	.. x .	
tigris					.. x ..
cerastes					.. x ..
mitchellii......					.. x ..
ruber				.. x ..	

The Desert Fauna.—The Colorado and Mojave Deserts with western and northern arms, one of which invades the southern part of the San Joaquin Valley, constitute the Californian portion of what may here be termed the Desert Fauna. This Fauna, as we have seen, is inhabited by thirty-one (or thirty-three) species and subspecies of reptiles, of which the following twenty-three (or twenty-four) occur in no other area of the State:

Gopherus agassizii,
Coleonyx variegatus,
Dipsosaurus dorsalis,
Uma notata,
Uma inornata,
Callisaurus ventralis,
Crotaphytus baileyi,
Crotaphytus wislizenii,
Sauromalus ater,
Uta graciosa,
Uta symmetrica,
Sceloporus magister,

Phrynosoma platyrhinos,
Phrynosoma m'callii,
Xantusia vigilis,
Cnemidophorus tigris,
Siagonodon humilis,
Chilomeniscus ephippicus,
Chionactis occipitalis,
Bascanion tæniatum (?),
Pituophis catenifer deserticola,
Crotalus tigris,
Crotalus cerastes,
Crotalus mitchellii.

This area shares with the southern coast or San Diegan Fauna alone only *Hypsiglena ochrorhynchus, Salvadora grahamiæ,* and probably *Arizona elegans,* and with the valleys or Californian Fauna possibly *Bascanion tœniatum,* while in common with both these areas it has *Uta stansburiana, Rhinocheilus lecontei, Bascanion flagellum frenatum, Thamnophis hammondii,* and perhaps *Sceloporus biseriatus.* Moreover, it lacks twenty-seven (or thirty-three) species and subspecies which occur in one or both of these adjoining Faunæ, and possesses none of those found in the Sierra Nevada and northern coast areas. The Desert Fauna is the most distinct of the minor life areas of California.

The San Diegan Fauna.—This area comprises the western portions or coastal slopes of San Diego, Riverside, Orange, San Bernardino, and Los Angeles Counties, excepting the higher lands, which belong rather with the Sierra Nevada. It is, in the main, a warmer and dryer area than the Californian Fauna, to which it is most closely allied.

The reptiles of this Fauna are twenty-eight (or thirty) in number, of which the following eight (or nine) are peculiar to it:

Uta mearnsi,	Cnemidophorus stejuegeri,
Sceloporus orcutti,	Verticaria hyperythra beldingi,
Phrynosoma blainvillii,	Lichanura roseofusca, †
Xantusia henshawi,	Crotalus ruber.
Xantusia riversiana, *	

It shares with only the Desert Fauna *Hypsiglena ochrorhynchus, Salvadora grahamiæ,* and probably *Arizona elegans;* with the Californian Fauna, *Anniella pulchra, Lampropeltis californiæ, Bascanion laterale,* and perhaps

---

* Insular.

† Occurs also near Tucson, Ariz.

Sceloporus biseriatus and *Lampropeltis boylii;* and with both these Faunæ, *Uta stansburiana, Rhinocheilus lecontei, Bascanion flagellum frenatum, Thamnophis hammondii,* and perhaps *Sceloporus biseriatus.* It lacks twenty-five species and subspecies of the Desert Fauna, and eight (or nine) of the Californian. Some species are common to it and to one or both of the northern areas—Sierra Nevadan and Pacific. The San Diegan Fauna is most closely allied to the Californian.

The Californian Fauna.—The Californian Fauna includes the western slope of the Sierra Nevada below the Sierra Nevadan Fauna, and extends thence westward to the ocean, excepting the area along the coast which constitutes the Pacific Fauna and that part of the San Joaquin Valley which belongs to the Desert Fauna. It appears to reach the coast in Ventura, Santa Barbara and San Luis Obispo Counties. Twenty-six (or twenty-nine) reptiles have been found within its limits. Of these, four are peculiar to it, as follows:

Crotaphytus silus,   Cnemidophorus tigris undulatus,
Phrynosoma frontale,   Tantilla eiseni.

It shares with the Desert Fauna alone possibly *Bascanion tæniatum;* with the Desert and San Diegan Faunæ, *Uta stansburiana, Rhinocheilus lecontei, Bascanion flagellum frenatum, Thamnophis hammondii,* and perhaps *Sceloporus biseriatus;* with the San Diegan Fauna alone, *Anniella pulchra, Lampropeltis californiæ, Bascanion laterale,* and perhaps *Sceloporus biseriatus* and *Lampropeltis boylii;* with the Pacific Fauna alone, *Sceloporus occidentalis;* with the Pacific and San Diegan Faunæ, *Clemmys marmorata* and *Gerrhonotus scincicauda;* and with all except the Desert Fauna, *Eumeces skiltonianus, Diadophis amabilis, Bascanion constrictor vetustum, Tham-*

nophis parietalis, Crotalus lucifer, and perhaps *Lampropeltis zonata* and *Lampropeltis boylii.* It lacks twenty-five (or twenty-seven) reptiles of the Desert Fauna, eleven (or twelve) of the San Diegan, two (or three) of the Pacific, and three (or five) of the Sierra Nevadan. The Californian Fauna is most closely allied to the San Diegan.

The Pacific Fauna.—The Pacific Fauna occupies a narrow strip along the coast as far south as Monterey County.* In the northern part of the State it widens and merges with the Sierra Nevadan, to which it is closely allied, but farther south it is confined to the western slope of the outer Coast Range. It is inhabited by fifteen (or seventeen) kinds of reptiles. Of these, *Gerrhonotus burnettii* and *Anniella nigra* are peculiar. *Sceloporus occidentalis* and *Contia mitis* it shares with the Californian Fauna only. *Lampropeltis zonata* is perhaps confined to this and the Sierra Nevadan Faunæ. Its other species are wide-ranging. It lacks thirteen (or sixteen) reptiles of the Californian Fauna, and four (or five) of the Sierra Nevadan. This Fauna is much better characterized by its batrachians, birds, and other animals than by its reptiles.

The Sierra Nevadan Fauna.—The fifth life area of California occupies a belt along the western (and eastern also?) side of the Sierra Nevada. One at least of its reptiles reappears in the mountains of San Diego and Riverside Counties. Twelve (or fourteen) kinds of reptiles have been taken in this area. *Gerrhonotus palmeri* and *Eumeces gilberti* are peculiar to it. *Sceloporus graciosus* occurs here but in no other part of California except the mountains of the southern part of the State.

* Probably to Santa Barbara in mountains.

Gerrhonotus principis perhaps is one of its inhabitants. *Lampropeltis zonata* may be peculiar to it and the Pacific Fauna. The other reptiles of the area are of rather wide distribution. It lacks seven (or eight) species of the Pacific Fauna, and seventeen (or twenty) of the Californian.

We have seen that there are five life areas in California* and that some of these are more closely allied than others. The Desert Fauna bears little resemblance to the San Diegan and Californian, and even less to the Pacific and Sierra Nevadan. Most of its species occur in western Arizona, southern Nevada, and northern Lower California.† It is, in fact, a part of the Lower Austral Zone of Merriam or South Warm Temperate of Allen. The San Diegan and Californian Faunæ have more in common. Apparently both belong to the Upper Austral Zone of Merriam, which is the Middle Warm Temperate of Allen. The Pacific and Sierra Nevadan Faunæ, also, are closely allied. They form a part of the Transition Zone of Merriam or North Warm Temperate of Allen, which, extending northward across western Oregon and Washington, forms another life area, which we may call the Puget Fauna.‡ It would seem then, that these three zones bend suddenly southward (irrespective of altitude) near the Pacific Coast. Thus it happens that their westernmost Faunæ lie north and south of each other instead of east and west — the Desert Fauna north of the San Lucan,§ the Californian north of the San Diegan, the Puget north of the Pacific and Sierra Nevadan.

Too little is known of the reptiles of Oregon, Wash-

---

* The colder portions of the mountains have not been considered.
† Except that part which belongs to the San Diegan Fauna.
‡ This is perhaps a part of the Canadian Zone.
§ The southern end of the peninsula of Lower California.

ington, Idaho, and Nevada to permit anything to be said of their distribution. However, lists of those which have been found in each State are given.

The following are the reptiles of Oregon:

Clemmys marmorata,
Crotaphytus wislizenii,
Uta stansburiana,
Sceloporus graciosus,
Sceloporus occidentalis,
Sceloporus biseriatus (?),
Phrynosoma douglassii,
Phrynosoma platyrhinos,
Gerrhonotus sciucicauda,
Gerrhonotus principis,
Eumeces skiltonianus,
Charina bottæ,

Contia mitis,
Diadophis amabilis,
Bascanion constrictor vetustum,
Bascanion tæniatum,
Pituophis catenifer,
Thamnophis parietalis,
Thamnophis parietalis pickeringii,
Thamnophis leptocephala,
Thamnophis vagrans,
Thamnophis vagrans biscutata,
Crotalus lucifer.

The following reptiles are known to inhabit Washington:

Clemmys marmorata,
Chrysemys bellii,
Sceloporus graciosus,
Sceloporus occidentalis,
Phrynosoma douglassii,
Gerrhonotus principis,
Charina bottæ,
Contia mitis,

Bascanion constrictor vetustum,
Pituophis catenifer,
Thamnophis parietalis,
Thamnophis parietalis pickeringii,
Thamnophis leptocephala,
Thamnophis vagrans,
Thamnophis vagrans biscutata,
Crotalus lucifer.

The reptiles of Idaho are:

Crotaphytus baileyi,
Crotaphytus wislizenii,
Uta stansburiana,
Sceloporus graciosus,
Sceloporus biseriatus,
Phrynosoma douglassii,
Phrynosoma platyrhinos,
Cnemidophorus tigris,

Charina bottæ,
Bascanion constrictor vetustum,
Bascanion tæniatum,
Pituophis catenifer,
Thamnophis parietalis,
Thamnophis vagrans,
Crotalus lucifer,
Crotalus confluentus.

The following species and subspecies have been found in Nevada:

Gopherus agassizii,	Heloderma suspectum,
Dipsosaurus dorsalis,	Cnemidophorus tigris,
Callisaurus ventralis,	Charina bottæ,
Holbrookia maculata approximans,	Lampropeltis boylii,
Crotaphytus baileyi,	Salvadora grahamiæ,
Crotaphytus wislizenii,	Bascanion flagellum frenatum,
Sauromalus ater,	Bascanion tæniatum,
Uta stansburiana,	Pituophis catenifer deserticola,
Uta graciosa,	Thamnophis parietalis,
Sceloporus graciosus,	Thamnophis vagrans,
Sceloporus biseriatus,	Crotalus lucifer,
Sceloporus magister,	Crotalus tigris,
Phrynosoma platyrhinos,	Crotalus cerastes.

Many kinds of reptiles vary so much that it is difficult to find two specimens which are alike in color and squamation. Sometimes the variations correspond with definite geographical areas, as in the case of *Cnemidophorus tigris* and *C. t. undulatus* or *Phrynosoma blainvillii* and *P. frontale*, but more frequently they are purely individual, as in *Charina* and many species of *Thamnophis*. Many reptiles are subject to chameleonic changes, or changes in accordance with the intensity of the light, or with the colors of objects by which they are surrounded. For these reasons the collector should strive to secure many specimens of each species.

Reptiles are to be found in all sorts of situations. The collector should study their habits if he would be successful in his search. Some kinds prefer moist places, while others are most abundant on barren hillsides or on the open desert. As a rule, reptiles like sunlight and warmth, but some species live in the thicker forests, and not a few are nocturnal.

Some reptiles may be caught with the hands unaided by any apparatus. Other species, too agile to be captured thus, may be secured by means of a slip-noose of horse-hair, wild oats, thread, or fine wire, deftly placed over the head of the victim and then tightened with a sudden jerk. However, by far the most satisfactory method of procuring reptiles is to shoot them. For this purpose small charges of fine shot are used in an auxiliary barrel, collecting pistol, or small caliber rifle. The last will prove much more effective if the rifling has been removed. When taken in the hands our reptiles often bite fiercely, but, even if they succeed in drawing blood, none except the rattlesnakes and the Gila Monster can cause any serious injury, for only these are poisonous.

Nothing is better for preserving reptiles than alcohol, though formalin may sometimes be used to advantage when little space is at the collector's disposal. Care should be taken to have the alcohol enter the body cavity, for if it does not do so the specimens will not be well preserved. The alcohol may be injected by means of a hypodermic syringe, or slits may be cut through the skin of the belly. These slits usually should be about half an inch long. One is ordinarily sufficient in case of a lizard, but in snakes several incisions should be made at interval of three or four inches. The specimens having been thus prepared, and labeled with the exact locality and date of collection, as well as with the collector's name and any notes upon habits, colors, etc., should be placed in strong alcohol. Care should be taken not to crowd the specimens into small jars with too little alcohol, for if this be done the reptiles will decay. If the number of jars at hand is so small as to necessitate crowding, the alcohol should be renewed each day until the specimens are thoroughly cured, after which only enough alcohol to cover them is needed.

The descriptions in the following pages are based
upon alcoholic specimens, except in a few instances
where it is distinctly stated that fresh specimens have
been used.   Alcohol does not preserve the colors of
reptiles well, so that living reptiles usually are more
brightly colored than the descriptions indicate.   In the
determination of colors Ridgway's "Nomenclature of
Colors" has been used as a guide.   Measurements are
given in millimeters, but may readily be converted into
inches by allowing twenty-five (25.4) millimeters to one
inch.   The tail is measured from the anus.   Limbs are
measured from the side of the body to the tip of the
longest toe, excluding the claw.   Many of the outline
figures of the heads, etc., of snakes are after Baird.
Other figures are original, having been drawn by Miss
Anna L. Brown.   I add here a glossary of some of the
terms used in works upon herpetology.

GLOSSARY.

Abdominal.—Pertaining to the lower surface of the
  body.

Abdominal plates.—Gastrosteges of snakes; the fourth
  pair of plastral plates of turtles.

Alveolar surface.—Masticatory surfaces just within the
  cutting edges of the jaws of turtles.

Anal plate.—The large scale just in front of the anus in
  most snakes, sometimes divided; one of the last
  pair of plastral plates.

Anteorbital.—See preocular.

Anterior.—Toward the head.

Antocular.—See preocular.

Antorbitar.—See preocular.

Anus.—The external opening of the cloaca.

Axilla.—The armpit.

Axillary.—Plates on the anterior surface of the bridge of turtles.

Azygous.—Single; not one of a pair.

Brachials.—Large scales on the arm.

Bridge.—That portion of the shell of a turtle which attaches the plastron to the carapace.

Canthus rostralis.—A slight continuation of the superciliary ridge separating the top from the side of the snout.

Carapace.—The upper portion of the shell of turtles.

Carinate.—Keeled.

Chin shields.—See geneials.

Cloaca.—A common chamber at the posterior ends of the alimentary and urogenital canals.

Collar.—Gular fold, especially of *Teiidæ*.

Costals.—The large plates on the sides of the carapace.

Dermal.—Pertaining to the skin.

Femoral pores.—Glands along the lower surface of the thigh.

Femorals.—Of turtles, the fifth pair of plastral plates; of lizards, plates on the thigh.

Frenal—See loreal.

Frontal—The large plate or plates on top of the head between the supraoculars. Sometimes applied to the prefrontals.

Frontoparietal.—Plates on top of the head between the parietals and the frontal.

Gastrosteges.—Large plates along the lower surface of the body in most snakes.

Gastrostiga.—See gastrosteges.

Geneials.—Large scales behind the mental of many snakes, often in two pairs—anterior and posterior.

Gular fold.—Transverse fold of skin of throat.

Gular plates.—The first pair of plastral plates.

Gulars.—Scales on throat.

Humerals.—The second pair of plastral plates.

Imbricate.—Lapped, as shingles.

Inferior.—Lower.

Infralabials.—Plates on the lower lip.

Inguinals.—Plates on the posterior surface of the bridge of turtles.

Internasals.—Scales on the top of the snout just behind the rostral plate.

Interparietal.—A plate on top of the head (of lizards) between the parietals and usually containing the pineal spot.

Juxtaposed.—Placed side by side, not imbricate.

Keel.—A ridge along a scale like the keel on an overturned boat.

Labials.—Plates on the lips; specially, on the upper lip.

Laterals.—Scales on the sides; the costals of turtles.

Loral.—See loreal.

Loreal.—In the space between the preoculars and nasals.

Maculate.—Marked.

Marginals.—The plates around the edge of the carapace.

Mental.—Same as symphyseal, but usually of snakes.

Mucronate.—Provided with a point or spine.

Nuchal plate.—The unpaired marginal plate of turtles on the median line at the front of the carapace.

Occipitals.—Plates behind the parietals. Sometimes applied to the parietals.

Parietals.—In most snakes, the largest and last plates on top of the head; in lizards, plates at the side of the interparietal and behind the frontoparietals.

Pectoral plates.—The third pair of plastral plates.

Plastral.—Pertaining to the plastron.

Plastron.—The lower portion of the shell of turtles.

Postabdominal.—Anal plate.

Postanal.—Behind the anus, especially a pair of large plates in the males of some lizards.

Posterior.—Toward the tail.

Postfrontals.—See prefrontals.

Postgeneials.—The posterior pair of geneials.

Postmentals.—Plates behind the mental. See sublabial and geneial.

Postocular.—Bounding the orbit behind.

Preanal.—In front of the anus.

Preanal pores.—Glands opening in front of the anus.

Prefrontal.—Scales between internasals and frontal. Sometimes applied to the internasals.

Pregeneials.—The anterior pair of geneials.

Prenasal.—Anterior nasal.

Preocular.—Bordering the orbit in front.

Pseudopreocular.—Small plate or plates below the preocular.

Reticulate.—Marked with lines like the meshes of a net.

Rostral.—Plate on the tip of the snout.

Scute.—A scale, especially a large flat one.

Subcaudals.—Urosteges.

Sublabials.—Plates below the infralabials.

Subocular.—Scales between the eye and the supralabials.

Superciliary.—Along the upper, outer edge of the orbit. Sometimes applied to the supraoculars of snakes.

Superior.—Upper.

Supracaudal.—Over the tail; the last pair of marginal plates of turtles, sometimes united.

Supralabials.—Upper labials. Also called superior labials or labials.

Supraocular.—Of snakes, the large scale over the eye; of lizards, the scales over the eye excepting the superciliaries.

Suture.—The line of joining.

Symphysal.—See symphyseal.

Symphyseal.—The scale on the tip of the lower jaw, especially of lizards. See mental.

Symphysial.—See symphyseal.

Urosteges.—Large scales on the lower surface of the tail in most snakes.

Vent.—The anus.

Ventrals.—Gastrosteges.

Vertebrals.—The large plates along the middle of the carapace.

Vertical.—Frontal.

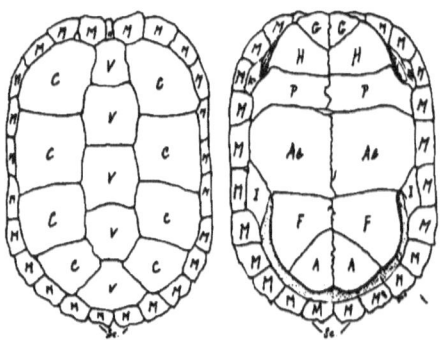

A.—Anal

Ab.—Abdominal.

Ax.—Axillary.

C.—Costal.

F.—Femoral.

G.—Gular.

H.—Humerals.

I.—Inguinal.

M.—Marginal.

N.—Nuchal.

P.—Pectoral.

Sc.—Supracaudal.

V.—Vertebral.

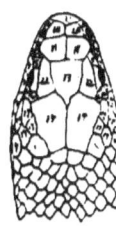 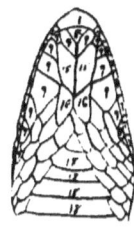 

1.—Rostral.
2.—Anterior nasal.
3.—Posterior nasal.
4.—Loreal.
5.—Preocular.
6.—Postoculars.
7.—Superior labials.
8.—Mental or symphyseal.
9.—Inferior labials.

10.—Internasals.
11.—Prefrontals.
12.—Supraoculars.
13.—Frontal.
14.—Parietals.
15.—Pregeneials.
16.—Postgeneials.
17.—Temporals.
18.—Gastrosteges.

Class REPTILIA.

The reptiles of the Pacific Coast and Great Basin belong to two great groups, to which they may be referred by the following

SYNOPSIS OF ORDERS.

a.—Body protected by a bony carapace or shell, covered with horny plates or leathery skin, jaws horny, without teeth. (Turtles.)
Testudines.—p. 28.

a².—Body not protected by a bony carapace; jaws provided with teeth.*
(Lizards and snakes)................... ......Squamata.—p.26. 3 8

Order I. TESTUDINES.

The *Testudinidæ* is, as yet, the only family of turtles known to be represented on the Pacific Coast and in the Great Basin. *Kinosternon* of the *Kinosternidæ*, however, lives in the Gila River of Arizona, and probably will be found in the Colorado as well. A species of *Trionychidæ* has been described† as having been taken in the Sacramento River, California. The skull of the type is missing, but in other respects the specimen appears to agree with the descriptions of a Chinese species *(sinensis)*. In view of this, and the additional fact that its describer afterward obtained several specimens of his supposed new species from Chinamen in San Francisco,‡ I cannot admit *"Pelodiscus californianus"* to be of Californian origin.

---

* The lower jaw only bears teeth in the *Leptotyphlopidæ.*
† See Rivers, Proc. Cal. Ac. Sci. (2), II. 1889. p. 233; Baur, Proc. Am. Philos. Soc., XXXI, 1893, pp. 218, 220.
‡ Dr. G. Baur has compared the skeleton of one of these specimens with that of *P. sinensis* and writes me that the two do not belong to the same species. Even admitting that the specimen which Rivers sent Dr. Baur is specifically identical with the type, I cannot admit that this turtle is indigenous to California until less questionable evidence of its occurrence here has been obtained.

Family I. TESTUDINIDÆ.

This widely distributed family contains a large number of turtles distinguished from others chiefly by osteological characters. The shell is firmly ossified, and covered with large horny plates of which eleven or twelve are on the plastron. The pectoral plates are in contact with the marginals. The latter are twenty-four or twenty-five in number. The neck can be completely drawn into the shell. Three genera are represented in the area under consideration.

SYNOPSIS OF GENERA.

a.—Feet not club-shaped, webbed; two supracaudal plates; skin on top of head not divided into scales.

  b.—Suture between abdominal plates less than twice length of suture between pectorals; inguinal plates not wedged in between abdominals and marginals . . . . **Clemmys.**—p. 29.

  b².—Suture between abdominal plates about twice length of suture between pectorals; inguinal plates wedged in between abdominals and marginals . . . . **Chrysemys.**—p. 32.

a².—Feet club-shaped, not webbed; one supracaudal plate; skin on top of head divided into scales . . . . **Gopherus.**—p. 35.

Genus I. CLEMMYS.

"*Clemmys*, WAGLER, Syst. Amph., 1830, p. 136 (type **caspica**);" "*Chelopus*, RAFIN., Atlant. Journ., 1832, p. 64;" "*Nanemys*, AGASSIZ, Contr. Nat. Hist. U. S., I, 1857, p. 442" (type **guttata**); "*Calemys*, AGASSIZ, l. c., p. 443" (type **muhlenbergii**); "*Glyptemys*, AGASSIZ, l. c., p. 443" (type **insculpta**); *Actinemys*, AGASSIZ, l. c., p. 444 (type **marmorata**); "*Mauremys*, GRAY, Proc. Zool. Soc. Lond., 1869, p. 500" (type **fuliginosa**); "*Sacalia*, GRAY, Suppl. Cat. Sh. Rept. Brit. Mus., 1870, p. 35;" "*Emmenia*, GRAY, l. c., p. 38;" "*Eryma*, GRAY, l. c., p. 44."

The shell is broad and low. The plastron is immovably united to the carapace by a broad bridge. There is no median ridge on the alveolar surface of the upper jaw parallel to the cutting edge. The internal openings of the nostrils are between the eyes. The fingers and toes are webbed. The skin on top of the head is not divided into scales. There are two supracaudal plates. The tail is moderate or long.

1.—Clemmys marmorata (Baird & Girard). PACIFIC
TERRAPIN.

Emys marmorata, B. & G., Proc. Ac. Nat. Sci. Phila., 1852, p. 177
(type locality **Puget Sound**).

Emys nigra, HALLOW., Proc. Ac. Nat. Sci. Phila., 1854, p. 91 (type
locality **Posa Creek, Lower** [=**Southern**] **California**); HAL-
LOW., U. S. Pac. R. R. Surv., X, 1859, Pt. IV, p. 3, pl. I.

Actinemys marmorata AGASSIZ, Contr. Nat. Hist. U. S., I, 1857, p.
444, II, pl. III, figs. 5-8; GIRARD, U. S. Explor. Exped., Herp.,
1858, p. 465, pl. XXXII.

Clemmys marmorata STRAUCH, Mem. Ac. St. Petersb. (7), V, No. 7,
1862, p. 108; BOULENGER, Cat. Chelonians Brit. Mus., 1889, p.
110; STEJNEGER, N. A. Fauna, No. 7, 1893, p. 162.

Clemmys Wosnessenskyi, STRAUCH, Mem. Ac. St. Petersb. (7), V,
No. 7, 1862, p. 114, pl.—— (type locality **Sacramento River,
California**).

Geoclemys marmorata, GRAY, Suppl. Cat. Shield Rept. Brit. Mus.,
1870, p. 27.

Chelopus marmoratus, COPE. Bull. U. S. Nat. Mus., No. 1, 1875, p.
53; YARROW, Bull. U. S. Nat. Mus., No. 24, 1882, p. 36.

Description.—Shell broad and low, broader posteriorly than anteriorly, more nearly round in young than in adults. Young with a median dorsal ridge not present in adults. Vertebrals five, broader than long. Costals four, first longest, second highest, last smallest. Nuchal very narrow. Marginals twelve on each side, supracau-dals being distinct. Plastron large, extending forward about as far as carapace, weakly notched posteriorly, truncate anteriorly. Its gular plates smallest, triangu-lar. Pectorals not much smaller than abdominals. Anals large, median suture between them longer than that of any other plastral plates. Bridge formed of pectoral and abdominal plates. Axillary and inguinal plates very small or absent. Head large, more or less flat topped, covered above and laterally with smooth skin. Upper jaw not hooked, sometimes notched at symphysis. Skin of neck and gular region granular. Limbs covered with scales; anterior with five, posterior with four, digits webbed to bases of long claws. Tail moderately long, tapering to tip, covered with scales in irregular whorls.

The coloration is very variable. In some specimens the carapace is olive or horn-color with few or no markings. In others a few broken and very irregular black lines are present. These lines frequently have become so numerous that, blending and crossing, they appear as the ground color, or form a very fine network through which the original ground color shows more or less indistinctly. Sometimes the carapace is almost black. The plastron is yellow, usually irregularly blotched with black or brown, or with dark lines along the posterior margins of the plates. The upper surface of the head may be unicolor or finely or coarsely reticulated with yellow and black. The chin and throat are yellow, often dotted with brown or black. The limbs and tail are yellow marked with black or brown, or brown marked with yellow. In young, the plates of the carapace show a central area of brown sometimes surrounded by a band of lighter brown or dull yellow, and the markings on the limbs, tail, neck, and gular region form irregular longitudinal bands.

Length of carapace 27	45	90	120	125	164
Length of plastron 24	40	79	109	117	153
Width of carapace................27	42	71	101	100	130
Width of plastron 21	33	57	85	85	105
Length of tail 21	30	36	38	48	65

Distribution.—The Pacific Terrapin probably occurs in all the fresh waters of the Pacific Slope from Lower

California to British Columbia. It has been taken at
both San Diego and Puget Sound, as well as at many
intermediate localities, some of which are: Mojave
River,* Posa Creek, South fork of Kern River near
Kernville, Mt. Diablo, Monterey, Santa Cruz, Peniten-
tia Creek, Coyote Creek, Palo Alto, San Francisquito
Creek, Los Gatos, San Francisco, San Rafael, Sonoma,
Nicasio, Sacramento River, McCloud River, Pitt River,
California; Klamath Falls, Oregon; Fort Steilacoom,
Washington.

Habits.—This is the terrapin of the San Francisco
markets, and is popularly known as the Mud Turtle or
Snapper. Very little is known of its habits. It is almost exclusively aquatic, preferring ponds and small
lakes to running water, but is sometimes encountered
on land while crossing from one body of water to another. It is sometimes caught with hook and line, and
probably is omnivorous. A specimen which I kept
alive laid three eggs in June and another in August.
The eggs are elliptical, with hard, white, limy shells,
and measure about thirty-four by twenty-one millime-
tres.

Genus 2.   CHRYSEMYS.

Chrysemys, GRAY, Cat. Tort., Croc., Amphis., Brit. Mus., 1844, p.
27 (types picta and bellii).

The shell is rather narrow, low or moderately high.
The plastron is immovably united to the carapace by a
broad bridge. There is a ridge on the alveolar surface
of the upper jaw parallel to the cutting edge. The internal openings of the nostrils are between the eyes.
The fingers and toes are fully webbed. The skin on top
of the head is not divided into scales. There are two
supracaudal plates. The tail is short or moderate.

---

*Cooper, Proc. Cal. Ac. Sci., II, 1863, p. 121.

2.—Chrysemys bellii Gray. WESTERN PAINTED TUR-
TLE.

Emys Bellii, GRAY, Syn. Rept. Griffith's An. Kingd., 1831, p. 31 (type
locality America?); DUM. & BIBR., Erpét. Générale, II, 1835,
p. 302; GRAY, Cat. Tort. Croc. Amphis. Brit. Mus., 1844, p. 27.
Emys Oregoniensis, HARLAN, Am. Journ. Sci. Arts, XXXI, 1837, p.
382, pl. —— (type locality ponds near Columbia River); HOL-
BROOK, N. A. Herp., I, 1842, p. 107, pl. XVI.
Chrysemys bellii, GRAY, Cat. Shield Rept., I, 1855, p. 33; AGASSIZ,
Contr. Nat. Hist. U. S., 1857, I, p. 439, II, pl. VI, figs. 8, 9.
Chrysemys oregonensis, AGASSIZ, Contr. Nat. Hist. U. S., 1857, I,
p. 440, II, pl. III, figs. 1-3.
Chrysemys nuttalii, AGASSIZ, Contr. Nat. Hist. U. S., 1857, II, p.
642 (new name for *C. oregonensis*).
Clemmys oregoniensis, STRAUCH, Mem. Ac. St. Petersb. (7), V, No.
7, 1862, p. 114.
Chrysemys cinerea var. *bellii*, BOULENGER, Cat. Chelonians Brit. Mus.,
1889, p. 74.

Description.—Shell comparatively narrow, depressed
but not very low, without dorsal keel. Vertebrals five,
usually longer than broad. Costals four, first longest,
second highest, last smallest. Nuchal very narrow.
Marginals twelve on each side, supracaudals being distinct. Plastron large, extending forward about as far
as carapace, weakly notched posteriorly, truncate or
rounded anteriorly. Gular plates smallest, triangular.
Pectorals very much smaller than abdominals; latter
longest, and with longest median suture. Axillary and
inguinal plates well developed, inguinal wedged in between abdominal and marginals. Head moderately
large, covered above and laterally, except sometimes on
temples, with smooth skin. Upper jaw not hooked,
sometimes notched at symphysis. Skin of neck and
gular regions granular or tubercular. Limbs covered
with scales, anterior with five, posterior with four, digits webbed to bases of long claws. Tail moderately long
or short. (Figure, p. 26.)

The carapace is indefinitely marbled above with olive, yellow, and dark brown. Near the middle of each costal plate is a vertical bar of yellow, while a narrow or indistinct yellow line runs along near the anterior margin of each costal plate. In front of each of these yellow markings is a large vertical blotch of dark brown. The vertebrals show traces of yellow longitudinal lines near their lateral edges. The edge of the carapace is yellow; from it yellow bars run up on the middles of the marginal plates. Between these yellow bars are dark brown or black ocelli with indefinite yellow concentric lines. The lower surfaces of the marginals are blotched with dark brown. The plastron is yellow, often heavily blotched with dark brown. The head and limbs are grayish or brownish olive, with numerous longitudinal yellow lines. There is a large, elongate, yellow or orange blotch behind the eye.

The colors of a living specimen were as follows: Large postocular blotch, scarlet-vermilion or flame scarlet; lines on side of head greenish white; lines on top of head dull yellow; eye Paris green with black cross-bar; costal and marginal markings chrome yellow and dark seal brown (almost black); costal bars tinged with cadmium orange.

Length of carapace 250
Length of plastron 232
Width of carapace 175
Width of plastron 139
Length of tail 32

Distribution.— Boulenger records specimens of this turtle as having been collected in British Columbia and at Walla Walla, "British Columbia" [=Washington?]. Harlan's "*Emys Oregoniensis*" was secured in ponds near the Columbia River. This turtle has twice been found in the San Francisco markets. In each instance the market-men told me that the turtle had been sent in with fish from the San Joaquin River near Stockton, California, but, when questioned, could not state positively that the lot had not come from Oregon or Washington.

Genus 3. GOPHERUS.

Xerobates, AGASSIZ, Contr. Nat. Hist. U. S., I, 1857, p. 446 (types polyphemus and berlandieri).

The shell is very broad and high. The plastron is immovably united to the carapace by a broad bridge. There is a ridge along the middle of the alveolar surface of each side of the upper jaw parallel to the cutting edge, except in front, where there is a longitudinal ridge at the symphysis. The internal openings of the nostrils are between the eyes. The limbs are club-shaped, the fore limbs flattened, without webs. The skin on top of the head is divided into scales. There is but one supracaudal plate.

3.—Gopherus agassizii (Cooper).   DESERT TORTOISE.

Xerobates agassizii, COOPER, Proc. Cal. Ac. Sci., II, 1863, p. 120 (type locality **Mountains of California near Fort Mojave**); TRUE, Proc. U. S. Nat. Mus., IV, 1881, p. 437.

Testudo agassizii, BOULENGER, Cat. Chelonians Brit. Mus., 1889, p. 156.

Gopherus agassizii, STEJNEGER, N. A. Fauna, No. 7, 1893, p. 161.

Description.—Shell broad and deep, often flattened above, its margin serrate all around, except in worn specimens, and usually more or less rolled upward over limbs. Growth-center of each plate smooth, but usually surrounded by beautifully ribbed shell. Vertebrals five, last largest and widest. Costals four, first longest, second and third about equally high, last smallest. Nuchal not much narrower than long. Marginals eleven and a half on each side, last pair being united to form a single supracaudal plate. Plastron large, extending forward beyond the carapace, notched posteriorly and sometimes anteriorly. Gular plates smallest, sometimes united, covering a narrow process of the plastron, which may be level or curved upward. Pectorals very much smaller than abdominals, with shortest median suture, except sometimes that of anals. Abdominals largest, with longest median suture. Humerals larger than femorals. Anals little longer than gulars. Axillary and inguinal plates well developed, latter varying from two to six, not extensively wedged in between abdominals and marginals. Head rather elongate, not very wide, covered above with flat scales larger anteriorly than posteriorly. Upper jaw not hooked, margins nearly straight, irregularly but finely serrate. Skin of neck with flattened granules. Anterior limbs large, heavy, much expanded laterally, covered in front and externally with large, hard, smooth scales, and provided with five stout claws.* Posterior limbs not compressed, covered around the edge of the circular sole with large scales, and provided with, four stout claws. Tail very short, slender distally.

The carapace is brown or horn-color, usually relieved, especially near the centers of the plates, with yellow. The head and limbs are brown. The plastron is yellow, shaded with brown along the edges of its plates.

Length of carapace.	215	260	285	310
Length of plastron.	210	265	285	300
Width of carapace.	160	212	230	240
Width of plastron	148	184	210	209

Distribution.—The Desert Tortoise is known only from the desert portions of southeastern California, southern Nevada, and probably Arizona.† It has been

---

* In one specimen the outer three claws of the right foot are united.
† Cox, Am. Nat., XV, 1881, p. 1003.

recorded from Yuma, Solado Valley, Leach Point Valley, Mountains near Fort Mojave, between Daggett and Pilot Knob, California; from Pahrump Valley and the Bend of the Colorado River, Nevada; and from Tucson, Arizona.* It occurs also at Needles and at Crater Summit, California.

Habits.—Almost nothing is known of the habits of this turtle if we except the following note by Mr. E. T. Cox:*

" This fellow is found on the basaltic mountains in the most arid parts of this dry country. He is a vegetarian, feeding, as I am told, on cacti. His flesh is highly esteemed as food by the Indians and Mexicans. You will perceive that his mandibles are notched or toothed. His legs are covered with bony scales, and his front toe nails are made long and strong for digging amongst the rocks, while the hind feet are round like an elephant's.

" When molested he draws in his head and closes the aperture with his legs by bringing the knees together in front of the head; the hind legs are also drawn in until the posterior spaces are closed by the feet, and in this way all vulnerable points are protected by impenetrable armor. In preparing the specimen, I found on each side, between the flesh and carapax, a large membranous sack filled with clear water; I judged that about a pint run out, though the animal had been some days in captivity and without water before coming into my possession. Here then is the secret of his living in such a dry region; he carries his supply of water in two tanks. The thirsty traveler, falling in with one of these tortoises and aware of this fact, need have no fear of dying for immediate want of water."

* Cox, l. c.

Order II. SQUAMATA.

The order *Squamata* contains the lizards and the snakes, which are regarded as constituting two suborders—*Sauri* and *Serpentes*. These suborders are very closely allied, and for convenience are treated together in the following

SYNOPSIS OF FAMILIES.

a.—Limbs well developed, pentadactyle.

   b.—Eye with movable lids.

   c.—Pupil elliptical, vertical; skin of top of head soft, free from skull, and covered with minute granules which are not appreciably larger than those on the back.

   Eublepharidæ.—p. 39.

   c^2.—Pupil round; top of head with plates or scales, not movable.

   d.—A series of femoral pores.

   e.—Lateral scales not abruptly smaller than ventrals; ventrals in numerous series; tongue not deeply divided at tip.

   Iguanidæ.—p. 42.

   e^2.—Lateral scales granular like dorsals, abruptly smaller than ventrals; ventrals in eight longitudinal series; tongue ending in two long slender points.

   Teiidæ.—p. 132.

   d^2.—No femoral pores.

   f.—Lateral scales very much smaller than dorsals and ventrals, usually hidden by a lateral fold; dorsal scales keeled.

   Anguidæ.—p. 101.

   f^2.—Lateral scales not much smaller than dorsals and ventrals; no lateral fold; scales smooth.

   g.—Scales on body flat, thin, and imbricate.

   Scincidæ.—p. 143.

   g^2.—Scales on body wart-like tubercles, usually bony, separated by narrow granular spaces.

   Helodermatidæ.—p. 120.

   b^2.—Eye without lids; pupil elliptical.....**Xantusiidæ.**—p. 122.

a^2.—Limbs absent (or rudimentary in *Boidæ*).

   h.—Ventral scales less than twice as broad as dorsals.

   i.—Plates on top of head much larger than those on body; anus bordered in front by several scales; no spine at end of tail.

   Anniellidæ.—p. 115.

   i^2.—Plates on top of head not larger than those on body; anus bordered in front by a single plate; a small spine at end of tail...........................**Leptotyphlopidæ.**—p. 150.

h².—Ventral plates more than twice as broad as dorsal scales.

j.—No rattle at end of tail; no pit between nostril and eye.

   k.—A small spur at each side of the anus; tail short and truncate, or top of head with small scales; pupil vertical.

                Boidæ.—p. 152.

   k².—No spur at side of anus; tail tapering; top of head with large plates; pupil round or elliptical.

     No enlarged fangs at front of mouth; coloration, if in rings, not red separated from black by white (yellow).

                Colubridæ.—p. 157.

j².—A horny rattle at end of tail; a pit between nostril and eye; a pair of large erectile fangs............**Crotalidæ.**—p. 214.

Suborder I. SAURI—Lizards.

Family II. EUBLEPHARIDÆ.

The members of this family are most closely related to the *Geckonidæ* or true geckos from which they are distinguished by the procœlian vertebræ and united parietal bones. The clavicle is dilated and loop-shaped proximally. The limbs are slender and the claws wholly or partially retractile into a sheath composed of two lateral plates whose superior edges are covered by a third. The eyes are rather large, with movable lids, and vertically elongate pupil.

Genus 4. COLEONYX.

Coleonyx, GRAY, Ann. & Mag. Nat. Hist., XVI, 1845, p. 162 (type **elegans**); " *Brachydactylus*, PETERS, Mon. Berl. Ac., 1863, p. 41 " (type mitratus).

In this genus the lower surface of each digit is provided with a series of small transverse plates. There are no enlarged chin shields behind the symphyseal plate. The skin is very soft, finely granular, and not attached to the bones of the skull. A small ear-opening is present. Males have a few preanal pores.

4.—Coleonyx variegatus (Baird).   BANDED GECKO.

Stenodactylus variegatus, BAIRD, Proc. Acad. Nat. Sci. Phila., 1858,
p. 254 (type locality **Colorado Desert**); *and* Mex. Bound Surv.
Rept., II, pl. XXIII, figs. 9–27.
Eublepharis fasciatus, BOULENGER, Cat. Liz. Brit. Mus., 1885, I, p.
234 (type locality **Ventanas, Mexico**).

Description.—Snout narrow but rounded and a little
longer than distance between orbit and ear-opening.
Head and upper surfaces of body covered with minute
granules slightly larger on snout than elsewhere. Rostral plate somewhat broader than high, and presenting
five edges. Behind it the slender prenasals, meeting on
the median line. A small supranasal plate. Symphyseal large, longer than wide. Six to eight upper and
as many lower labials, decreasing in size posteriorly.
Eyelids bearing a fringe of pointed scales. Ear-opening
small, oval, and oblique. Feet, belly, and tail covered
with small, smooth, imbricate scales. Digits short.
Tail conical, about as long as head and body. A small
spur on each side of tail near its base. Males with a
short series of six to eight preanal pores.

The back is crossed by about five broad bands of dull
brown between which are narrower wavy bands of white.
A white horseshoe-shaped line on the neck passes just
above the ears and ends near the eyes. The head is
brown, or whitish with irregular brown spots. A dark
brown band runs from the eye to the nostril. The la-

bials are spotted with brown and white. The tail is cross-barred with the colors of the back, but the white areas are often partly occupied by brown spots. One specimen has the brown bands of the back narrower than the white ones. The lower surfaces are white.

A living specimen of *Coleonyx variegatus* is colored as follows: Across the back are five wide bands of dark walnut brown, palest centrally, and separated from one another by dull Naples yellow bands of about half their width. The tail is similarly cross-banded. The upper surfaces of the head and limbs are fawn color, the limbs being faintly and the head strongly marked with small irregular spots of walnut brown. The edges of the eyelids are white. A white line runs back from the eye to the top of the neck, where it meets or almost meets its fellow rf the opposite side. A walnut line, bordered above and below with white, connects the eye and nostril. The tongue is rich pink with a bright red tip. The lower surfaces are white. The eye is pale grayish yellow with a network of fine black lines.

Length to anus.......... 32	57	61	65
Snout to orbit................................... 3	5	5	6
Snout to ear............... 8	13	13	14
Orbit to ear 3	5	5	6
Fore limb..12	19	22	23
Hind limb..16	27	28	28
Base of fifth to end of fourth toe................... 4	6	7	8

Distribution.— The Banded Gecko probably ranges over the greater part of the Mojave and Colorado Deserts of southeastern California. In the north, it has been taken in Owen's and Death Valleys in Inyo County; in the west, at Mojave in Kern County and San Jacinto in Riverside County; and in the southeast, at Fort Yuma in San Diego County. Thence its range extends east to Tucson, Arizona, and south into Mexico.

Habits.—Very little is known of the habits of this lizard. An individual kept in confinement for more than a year spent most of his time in a hole provided in the ground of his cage. His food during this period consisted entirely of houseflies. His usual time of feeding was after dark, but not infrequently he would snap up a fly which chanced to stray into the mouth of his burrow during the day, and sometimes would come forth in search of prey while the sun was shining brightly on his den. When stalking flies, his movements were so slow as almost to be imperceptible until he was within range and could seize the coveted morsel with one instantaneous snap. If blown upon, he would raise himself and stand with legs straight and rigid. When first sent to me, this lizard had the skin of the occiput raised into a large hood, but whether this was a nuptial ornament or due to some accident I cannot tell.

Family III.   IGUANIDÆ.

The members of this family present, in their strange diversity of form, a series of pleurodont lizards which closely parallels in the New World the acrodont *Agamidæ* of the Old. The *Iguanidæ* are diurnal lizards having eyes with round pupils and well developed lids. The tongue is short, thick, and but slightly notched anteriorly. Femoral pores are present in North American species. The clavicle is not dilated, except in the Central American *Basiliscinæ*. Some species of *Sceloporus* and *Phrynosoma* are said to be ovoviviparous. Californian Iguanians may be distinguished by the following

SYNOPSIS OF GENERA.

a.—A low dorsal crest composed of one longitudinal series of enlarged scales..**Dipsosaurus.**—p. 43.

a².—No dorsal crest.

  b.—Head without spines.

    c.—One or more well developed transverse gular folds.

      d.—Toes fringed laterally with prominent movable spines.

                        Uma.—p. 46.

      d².—Toes without spines.

        e.—Supralabials strongly imbricate; symphyseal plate smaller than largest infralabial.

          f.—An ear-opening.....................**Callisaurus.**—p. 47.

          f².—No ear-opening...................**Holbrookia.**—p. 51.

        e².—Supralabials not imbricate; symphyseal plate not smaller than largest infralabial.

          g.—No large interparietal plate; caudal scales small, not strongly keeled nor sharply pointed.

            h.—Ear without strong denticulation and neck without spinose tubercles; superciliaries imbricate; tail long and tapering...................**Crotaphytus.**—p. 53.

            h².—Ear with strong denticulation and neck with numerous spinose tubercles on lateral folds; superciliaries not imbricate; tail scarcely longer than distance from snout to vent..................**Sauromalus.**—p. 60.

          g².—A very large interparietal plate; caudal scales large, strongly keeled, and sharply pointed.

                    Uta.—p. 63.

    c².—No complete transverse gular fold.........**Sceloporus.**—p. 73.

  b².—Head with large spines posteriorly..........**Phrynosoma.**—p. 89.

Genus 5.   DIPSOSAURUS.

Dipsosaurus, HALLOW., Proc. Ac. Nat. Sci. Phila., 1854, p. 92 (type **dorsalis**).

The scales of the median dorsal row are slightly enlarged, forming a small crest. The head is covered with small convex subgranular plates. The dorsal and caudal scales are small. There is one strong transverse gular fold. Femoral pores are numerous. Males do not have enlarged postanal plates. Digits each have a series of keeled plates below.

5.   **Dipsosaurus dorsalis** (Baird & Girard).   Crested
Lizard.

Crotaphytus dorsalis, Baird & Girard, Proc. Acad. Nat. Sci. Phila.,
1852, p. 126 (type locality **Desert of Colorado, Cal.**).
Dipsosaurus dorsalis, Baird, U. S. Mex. Bound. Surv., III, Rept.,
pl. XXXII, figs. 7–13.

Description.—Head short, rounded, and rather high.
Nostril opening laterally in a single rounded plate which
is separated from the large rostral by one or two rows of
granules.   Supraocular regions separated from each
other by two or three series of small convex plates and
covered with very small plates and granules.   A large
subocular, followed and preceded by several smaller
ones.   A series of long, strongly imbricate superciliaries.   Labials small, about equal in size, and from eight

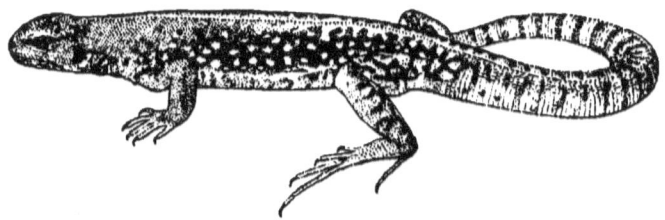

to eleven in number in each series.   Symphyseal plate
nearly triangular and forming the base of a V-shaped
series of slightly enlarged plates.   Gulars small, either
convex or flattened.   Eyelids very slightly fringed.
Ear-opening very large, almost vertical, and with a
very weak anterior denticulation.   Dorsal crest composed of slightly enlarged, strongly keeled scales.
Other dorsals small, keeled, juxtaposed, and in series
which converge toward the dorsal line posteriorly.
Ventrals larger than dorsals and caudals, smooth and
imbricate.   Sides covered with small granular scales.
Tail long, tapering, slightly crested, and with whorls of

obliquely keeled scales. Limbs rather long, covered with keeled scales and granules. Femoral pores varying from sixteen to twenty-five in number.

The general color is grayish brown above, variously barred and reticulated with dark brown and slate, and spotted or blotched with light gray or white. These markings are often less distinct near the vertebral line than laterally. The upper surface of the head is grayish, brownish, or yellowish, more or less clouded with slate, darkest on the supraocular regions. The tail is whitish, yellowish, grayish, or brownish, marked with rings of brown or slate. The lower surfaces are white, marked on the chin and gular region with longitudinal or oblique lines of brown or bluish gray.

The following color description was taken from a fresh male shot at Yuma, Arizona, October 1, 1894: The head is creamy, tinged on the sides with vinaceous and on the supraocular regions with black; below, white with indistinct gray markings. The back is cream with numerous transverse gray bars and more or less broken longitudinal lines of dull Chinese orange. These lines become spots on the sides. The tail is half-ringed with more or less connected spots of the same orange color. The belly is white with a large patch of reddish orange on each side.

Length to anus 47	73	94	105	126	133
Length of tail 91	151	172	190	232	255
Snout to orbit................... 4	5	6	7	8	9
Snout to ear..................... 10	15	18	19	21	23
Orbit to ear..................... 3	4 ·	5	6	6	7
Fore limb....................... 20	29	38	40	43	54
Hind limb 37	55	68	77	81	95
Base of fifth to end of fourth toe.. 17	21	29	32	34	39

Distribution.—In California, the Crested Lizard ranges over the lower levels of the Colorado and Mojave Deserts, pushing its way north to Owen's, Panamint, Death, and Amargosa Valleys. West of the desert region it has not been found and doubtless does not occur. It is quite common at Yuma. In Nevada, it has been taken on the Amargosa Desert and at Callville on the Great Bend of the Colorado River.

Habits.—At Yuma, this lizard lives in burrows in the mounds of sand which the winds heap up around the cactus bunches; the spines of the cactus serving to pro-tect them from the quick swoops of hungry hawks and the digging of larger enemies. Dr. C. Hart Merriam says:* "It is a strict vegetarian, feeding on buds and flowers, which it devours in large quantities. No in-sects were found in any of the stomachs examined; some contained beautiful bouquets of the yellow blos-soms of acacia, the orange malvastrum, the rich purple Dalea, and the mesquite *(Prosopis juliflora);* others con-tained leaves only."

Genus 6.   UMA.

Uma, BAIRD, Proc. Ac. Nat. Sci. Phila., 1858, p. 253 (type **notata**).

"A series of elongate free scales on each side of the digits, and on the external side of the sole." "Ears distinct. A very long infraorbital plate. Palate with-out teeth. Outer face of upper labials plane and broad-ly vertical; the labials themselves much imbricated, and very oblique. Scales of body above equal, much smaller than ventral ones. Interorbital space with two series of plates. Claws very long, slender and straight."

SYNOPSIS OF SPECIES.

a.—Nine rows of loreal plates; black spots on side of belly.

U. notata.—p. 47.

a².—Five or six rows of loreal plates; no black spots on side of belly.

U. inornata.—p. 47.

---

* N. A. Fauna No. 7, 1893, p. 105.   .ˑ

6.—Uma notata Baird. SAND LIZARD.

Uma notata, BAIRD, Proc. Acad. Nat. Sci. Phila., 1858, p. 253 (type locality **Mojave Desert**); COPE, Am. Nat., 1895, p. 939.

Description.—"Labial scales weakly keeled; nine loreal rows; fourteen supraorbital rows; hind foot longer, two-fifths head and body; femoral pores nineteen. Black spots on side of belly but no crescents on throat." "Head about two-fifths the head and body. Above light pea green, spotted with darker green. Beneath white."

Distribution.—Mojave Desert.

Habits.—Unknown.

7.—Uma inornata Cope. PLAIN SAND LIZARD.

Uma inornata COPE, Am. Nat. 1895, p. 939 (type locality **Colorado Desert, San Diego Co., Cal.**).

Description.—"Labial scales strongly keeled; five or six loreal rows; ten or eleven supraocular rows; hindfoot shorter, one-third head and body; femoral pores nineteen. No black spots on belly or crescents on throat."

I do not feel certain that these two supposed species are really distinct. The slight differential characters which have been pointed out are manifestly rather variable, and each name is based upon a single specimen, at least one of which *(U. notata)* is young and in a very poor state of preservation.

Distribution.—"Colorado Desert, San Diego County, California."

Genus 7. CALLISAURUS.

Callisaurus BLAINV., Nouv. Ann. Mus., IV, 1835, p. 286 (type draconoides); *Megadactylus* FITZINGER, Syst. Rept., 1843, p. 59 (type draconoides); *Homalosaurus* HALLOW., Proc. Ac. Nat. Sci. Phila., 1852, p. 179 (type ventralis).

The lizards of this genus have the body and tail considerably flattened, legs very long, and the head rounded

when seen from above but pointed in profile. The head
is covered with irregular plates, the largest of which is
the interparietal. The labials are produced laterally and
are strongly imbricate. The ear-opening is large. The
dorsal scales are very small and nearly uniform. There
are no fringes of movable scales on the digits. Long
series of femoral pores are present. There are two or
more transverse gular folds. Males have enlarged postanal plates.

8.—Callisaurus ventralis (Hallowell). GRIDIRON-TAILED
LIZARD.

> *Homalosaurus ventralis* HALLOW., Proc. Acad. Nat. Sci. Phila., VI,
> 1852, p. 179 (type locality **New Mexico**); and Sitgr. Zuni &
> Colorado Rivers, 1853, p. 117, pl. 6.

Description.—Head rather short and low, with well
developed canthus rostralis. Nostrils large, opening
on upper surface of snout. Supraocular regions covered
with small plates and separated from each other by one

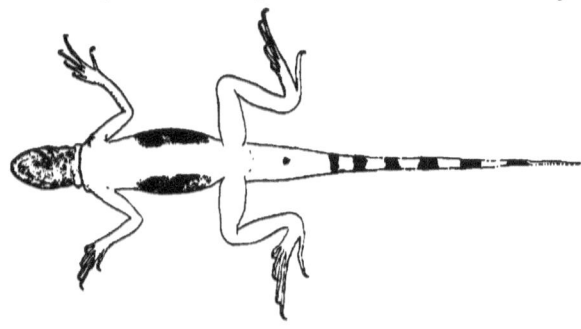

or two rows of slightly larger plates. Upper head plates
(except interparietal) small and irregular, largest on
frontal and prefrontal regions, everywhere smooth and
flat. Several subocular plates, middle one very long and
strongly keeled. Superciliaries rather small, but strongly

imbricate. Eyelids bearing a well developed fringe. Supralabials strongly imbricate and produced laterally so as to form a series of curves when seen from above. Infralabials small and juxtaposed. Below them, several series of flat sublabial plates. Gulars granular and smooth, growing larger and imbricate on posterior fold. Back and sides covered with small flattened granules, which change gradually into much larger smooth ventrals. A dermal fold usually extending along each side between limbs. Tail of moderate length, much flattened. Its scales slightly imbricate, and along its edge, pointed. Limbs very long and slender. Ear-opening large, without denticulation. Femoral pores varying from fourteen to eighteen.

The general color above is grayish, dotted and spotted with white or pale gray, and with indications of dark dorsal blotches which are most distinct in females and young. The top of the head is rich cream, clouded with dark slaty gray. The upper surfaces of the limbs and tail are crossed by more or less undulating bands of dark brown or blackish slate. A dark line, bordered above and below with white, runs along the back of the thigh. The throat is white, more or less clouded with gray. The lower surface of the tail is white with about seven cross-bars of intense black. The belly is whitish. Males have a large blue patch, marked with two oblique wedge-shaped black blotches, on each side.

The following color description was taken from a fresh male shot at Yuma, Arizona, October 1, 1894: The top of the head is cream; the upper surface of the fore limbs bright lemon yellow; the hind limbs slightly tinged with yellow; neck and fore back pale gray spotted with lighter; back like neck, but suffused with bright lemon yellow, which extends down over the sides and changes to orange

near the large verdigris green blotches on the sides of
the belly. There is a reddish orange area in front of
each of these green blotches. The throat is gray with
a half-concealed vermilion spot.

Length to anus................44	72	74	82	86	88
Length of tail....................59	102	98	107	117	
Snout to back of interparietal....... 9	13	13	14	15	15
Snout to ear...... 10	15	14	16	16	16
Width of head............. 9	13	13	14	14	14
Fore limb....................24	42	41	45	49	46
Hind limb 41	70	65	76	79	
Base of fifth to end of fourth toe...17	31	30	33	35	

Distribution.—The range of the Gridiron-tailed Lizard
in California is, in general, the whole southeastern part
of the state comprised in the Mojave and Colorado Deserts. In the north, it extends to Owen's, Saline, and
Death Valleys, in Inyo County; in the west, to the lower
parts of Antelope Valley in Los Angeles County, and
the top of San Gorgonio Pass near Banning, Riverside
County. The species has been found also at Fish
Springs and in a little 'island' of the desert at Oak.,
Grove, San Diego County, and is very abundant along
the Colorado River. "Southern and western Nevada
as far north as Pyramid Lake" are also occupied by it.

Habits.—The Gridiron-tailed Lizard "inhabits the
open deserts and runs with great swiftness over the sand
and gravel beds, carrying its tail curled over its back as
if afraid to let it touch the hot surface of the earth. It
starts off at full speed as if fired from a cannon, and
stops with equal suddenness, thus escaping or eluding
its enemies, the coyotes, hawks, and larger lizards.
When running it moves so swiftly that the eye has difficulty in following, and when at rest its colors harmonize
so well with those of the desert that it can hardly be

seen. This species feeds on insects and the blossoms and leaves of plants in about equal proportion." *

Genus 8. HOLBROOKIA.

Holbrookia, GIRARD, Proc. Am. Assoc. Adv. Sci., IV, 1851, p. 200 (type **maculata**); *Cophosaurus*, TROSCHEL, Arch. f. Nat., 1850 (1852), p. 389 (type **texanus**).

This genus contains a number of lizards similar to *Callisaurus* but with the ears hidden under the skin. The head, rounded when seen from above but pointed in profile, is covered with irregular plates, the largest of which is the interparietal. The labials are produced laterally and are strongly imbricate. There is no ear-opening. The dorsal scales are very small and nearly uniform. There are no fringes of movable spines on the digits. Long series of femoral pores are present, as are one strong and one or more weak gular folds. Males have enlarged postanal plates.

9.—**Holbrookia maculata approximans** (Baird). WESTERN EARLESS LIZARD.

Holbrookia approximans, BAIRD, Proc. Ac. Nat. Sci. Phila., 1858, p. 253 (type locality **Lower Rio Grande**).
Holbrookia maculata maculata, YARROW, Bull. U. S. Nat. Mus., No. 24, 1882, p̄, 49.
Holbrookia maculata var. *flavilenta*, COPE, Proc. Ac. Nat. Sci. Phila., 1883, p. 10 (type locality **Lake Valley, New Mexico**); STEJN-EGER, N. A. Fauna, No. 3, p. 109.

Description.—Head rather short and low. Nostrils large, opening on upper surface of snout. Supraocular regions covered with small plates or granules and separated from each other by one or two rows of slightly larger plates. Upper head-plates, except interparietal, small and irregular, largest on frontal and prefrontal regions, everywhere smooth and rather flat. Several

---

subocular plates, middle one very long and strongly
keeled. Superciliaries rather small, but strongly imbricate. Eyelids with well developed fringe. Supralabials
strongly imbricate and produced laterally so as to give
snout a rounded outline when seen from above. Infralabials small and juxtaposed. Several series of flat sublabials. Gulars granular and smooth, growing larger
and imbricate on posterior fold. Back and sides covered with scales or granules, largest near middle of
back and changing gradually to larger smooth ventrals.
A dermal fold usually present along each side between
limbs. Tail of moderate length, flattened near the
body. Its scales feebly keeled and slightly imbricate.
Limbs rather long, not very slender, eight to thirteen
femoral pores.

The color above is gray, yellow, or brown, with two
or four series of dark undulate blotches and numerous
light spots. The blotches are often more or less obsolete and are most distinct in females and young. The
top of the head is colored like the back, but without
definite markings. The limbs may be unicolor or
crossed by dark bars. The throat is white or yellow,
sometimes marbled with dusky. The belly is white or
yellow with two or three black bars on the sides usually
surrounded with blue. The tail is grayish or brownish
above, white or yellow below.

Length to anus................52	53	56	58	64	67
Length of tail................40	58	66	55	56	60
Snout to back of interparietal......... 9	9	10	10	11	11
Width of head................ 9	9	9	10	11	11
Fore limb................21	26	29	28	26	30
Hind limb 33	42	44	43	40	45
Base of fifth to end of fourth toe.......12	17	17	17	14	17

Distribution.—This *Holbrookia* is very common in
Arizona and has been taken at Dome Cañon, Nevada.

Yarrow has recorded a *Holbrookia* as having been collected at Fort Tejon, California, but there is not the slightest probability that this is correct.

Genus 9. CROTAPHYTUS.

Crotaphytus, HOLBROOK, N. A. Herpetology, II, 1842, p. 79 (type collaris).

The head and body are somewhat depressed, and much shorter than the tapering tail. All of the head-plates are small. The labials are not imbricate. The ear-opening is large, without strong denticulation. The dorsal scales are small and nearly uniform. Long series of femoral pores and one or more tranverse gular folds are present. There are no spinose tubercles on the neck. The superciliaries are imbricate. Males have enlarged postanal plates.

SYNOPSIS OF SPECIES.

a.—Two black bars across the shoulders..........C. baileyi.—p. 53.
a².—No black bars across shoulders.
   b.—Greatest width of head less than distance between nostril and ear-opening........................ ....C. wislizenii.—p. 56.
   b².—Greatest width of head equal to or greater than distance between nostril and ear-opening ....C. silus.—p. 59.

10.—**Crotaphytus baileyi** Stejneger. BAILEY'S LIZARD.

Crotaphytus baileyi, STEJN., N. A. Fauna, No. 3, 1890, p. 103, pl. XII, fig. 1 (type locality **Painted Desert, Little Colorado River, Arizona**); and l. c., No. 7, 1893, p. 165.

Description.—Head large, depressed, and very distinct from the neck on account of swollen temples. Its plates all small, but largest and somewhat convex on snout. Two longitudinal rows of shields separating supraocular regions. Nostrils large and opening laterally, each in a round plate nearer to end of snout than, to orbit. Superciliaries small but imbricate. Supralabials rather prominent and of nearly equal size. A large subocular plate.   Ear-opening large, oblique, with very slight anterior denticulation.   Supraoculars, temporals, and gulars subgranular.   Lower labials a little larger than upper, bordered below by several series of plates larger than gulars.   Symphyseal plate large, followed by a pair of large shields.   One or two gular folds, continued on sides of neck.   Back and sides covered with small granules, which pass gradually into larger, smooth, flat scales on belly.   Sides irregularly plicate. Tail tapering, nearly twice as long as head and body, and furnished with whorls of small, smooth plates. Femoral pores varying in number from sixteen to twenty-two in each series.   Males with enlarged postanal plates.

The general color is greenish, bluish, olive, grayish, or pale brown, variously dotted, blotched, reticulated, and cross-lined with pale gray or white.   Two parallel oblique bands of intense black or very dark brown cross the shoulders, but often do not meet on the nape.   The tail sometimes bears large brown spots.   The head is irregularly spotted and reticulated laterally and inferior-

ly. The throat and belly are white, more or less suffused
with blue; the latter sometimes with large brown lateral
blotches.

Dr. Stejneger has given* the following description of
the fresh colors of a young individual obtained near the
Little Colorado River, Arizona:—

"Head above pale sepia, inclining to clay color; anterior portion of upper neck in front of the first black
collar pale blue, with several longitudinal marks of
'coral red;' space between the two black collars pale
'oil green,' with a narrow transverse collar of coral red;
ground color of back dull oil green, fading posteriorly
on hind legs and tail to a grayish 'pea green,' the back
densely covered with rather large dark grayish olive
blotches, which only allow the ground color to show
through as a fine reticulation; the second black collar
bordered posteriorly with a wide line of 'lemon yellow,'
the back being crossed by five similar lines, fading posteriorly and more or less alternating on the lateral halves
of the body; tail with transverse bars of dark grayish
brown; fore legs above 'apple green,' nearly yellow on
hand and faintly barred with the latter color; under
surface pale greenish-white, palms slightly pinkish, tail
nearly white. Tongue deep pink; pharynx blackish
carmine; palate ultramarine blue. Iris brassy greenish-
yellow."

Length to anus	60	82	90	99	100	106
Length of tail	111	173	175	242	210	229
Snout to orbit	5	8	9	9	10	12
Snout to ear	16	23	25	30	29	32
Width of head	14	20	23	25	25	26
Fore limb	27	39	42	50	48	55
Hind limb	54	82	82	98	90	103
Base of fifth to end of fourth toe	20	29	29	33	31	37

* N. A. Fauna, No. 3, 1890, p. 105.

Distribution.—This also is a lizard of the desert, but seems not to live upon its lower levels, preferring the more mountainous portions between the altitudes of about 4,500 and 6,500 feet. In suitable localities it is quite abundant in Inyo, Kern, and San Bernardino Counties, and doubtless occurs also in the eastern parts of Riverside and San Diego Counties, California. It has been recorded from Idaho (mouth of Bruneau River) and is common in western Utah and in most parts of Nevada (Oasis Valley, Juniper Mts., Desert Mts., North Kingston Mts., Reno to Pyramid Lake, Dome Cañon, etc.).

11.—Crotaphytus wislizenii Baird & Girard. LEOPARD LIZARD.

Crotaphytus Wislizenii, B. & G., Proc. Acad. Nat. Sci. Phila., VI, 1852, p. 69 (type locality **Santa Fé, New Mexico**); *and* Stansbury's Exped. Gt. Salt Lake, 1853, p. 340, pl. III; BAIRD, Mex. Bound. Surv., 1859, pl. XXXI.

Crotaphytus Gambelii, BAIRD & GIRARD, Proc. Ac. Nat. Sci. Phila., VI, 1852, p. 126 (type locality **California**).

Description.—Head large, depressed, not so distinct from neck as in *C. baileyi*. Its plates all small but largest and somewhat convex on snout. Three to five longitudinal rows of shields separating supraocular regions. Nostril large and opening laterally in a round plate much nearer to end of snout than to orbit. Superciliaries small but imbricate. Rostral plate wide but very low. Supralabials of nearly equal size. A long subocular plate. Ear-opening large, oblique, with very slight anterior denticulation. Supraoculars and temporals granular, as also gulars. Lower labials slightly larger than upper, and bordered below by several series of small plates, larger than gulars. Symphyseal plate very large, but shields behind it not so large as in *C. baileyi*. From one to three transverse gular folds only one well developed. Back and sides covered with small granules, largest centrally and passing gradually into the larger scales on the belly. Latter imbricate and sometimes keeled. Irregular dermal folds usually present on sides. Tail conical, a little more than twice length of head and body, and covered with whorls of small scales. Femoral pores varying in number from about seventeen to twenty-three. Males with enlarged postanal plates.

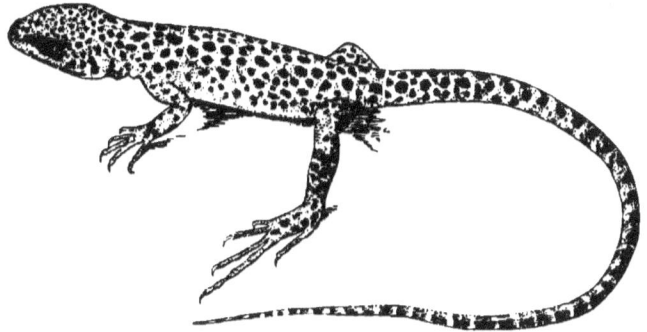

. In young specimens the head is dark brown above, with cream-colored lines surrounding the orbits and supraocular regions and running up the median line of the snout from the rostral plate. The back is grayish brown with white or cream-colored cross-lines, which may either meet or alternate, on the median line, with those of the opposite side. Between each pair of these cross-lines is a round spot of dark reddish brown. The tail is marked like the back, but not so regularly. The limbs are brown with irregular spots and lines of white. The lower surfaces are yellowish white, marked on the throat with longitudinal lines of dark brown. As the animals become larger the brown dorsal spots become smaller and more numerous, so that there are several between each pair of light cross-lines. The whole coloration becomes paler, as if faded, and the pattern less distinct. Usually the light cross-lines fade first, leaving the spots fairly distinct, but the reverse order of disappearance may occur. In some very old specimens the cross-lines have entirely vanished and the brown spots have become very minute. There is also a good deal of purely individual color variation.

During the breeding season some females have the under surfaces and sides of the tail and body suffused with deep salmon or salmon-red. This color disappears in alcohol.

Length to anus	47	69	87	89	107	119
Length of tail	77	160	175	204	225	
Snout to orbit	4	6	8	8	10	12
Nostril to ear	10	14	18	18	21	25
Width of head	9	12	16	16	19	22
Fore limb	19	28	34	34	43	46
Hind limb	32	54	69	74	80	87
Base of fifth to end of fourth toe	13	21	27	30	30	33

Distribution.—The Leopard Lizard occupies almost the entire desert region of eastern California, ranging from San Diego County across the Colorado and Mojave Deserts to the smaller desert valleys of the Great Basin, where it is quite common in Inyo County. Through Walker Pass, it reaches the western slope of the Sierra Nevada, where it occurs in Kern Valley. It has been collected also at San Jacinto on the western slope of the Coast Range of Riverside County. At Mojave Station, Kern County, this lizard is rather rare, but at Needles, San Bernardino County, it may readily be found on the sand banks of the Colorado River. It ranges across Nevada (Pyramid Lake, Panaca, Vegas Valley, Mt. Magruder, Quartz Spring, Amargosa Desert, Sarcobatus Flat, Charleston Mts., Grapevine Mts., Timpahute Mts.,

Indian Spring Valley, Pahrump Valley, Pahranagat Valley, Oasis Valley, Wadsworth, Dome Cañon, Truckee River) to Utah (St. George, etc.), Oregon (Dalles), and Idaho (along Snake River).

Habits.—"The leopard lizard is chiefly a vegetarian, feeding on the blossoms and leaves of plants; but is also carnivorous, devouring the smaller lizards, horned toads, and even its own kind, besides large numbers of insects, as determined by the examination of many stomachs."* It usually is sluggish, and may be captured without much difficulty.

12. — Crotaphytus silus Stejneger. SHORT-NOSED LEOPARD LIZARD.

Crotaphytus silus, STEJN., N. A. Fauna, No. 3, 1890, p. 105 (type locality **Fresno, Calif.**); *Id., ibid.,* No. 7, 1893, p. 170.

Description.—This species is very similar to *C. wislizenii,* but has a much shorter and more truncate snout. The greatest width of the head is equal to or greater than the distance from the nostril to the ear-opening. The distance between the nostril and the inner anterior orbital angle is considerably less than the vertical diameter of the ear-opening.

"The coloration is also essentially different. In *C. silus* the rounded dorsal spots are larger, especially the two median rows, so that of the latter there is only one longitudinal series between the light cross-bands. The latter are very broad and distinct and do not seem to disappear as the animal grows larger. In some specimens the interspaces between the light bands are solidly dark, the spots indicated only by somewhat ill-defined patches of saturated ferrugineous."

Distribution.—"This species seems to be closely re-

---

* N. A. Fauna, No. 7, 1893, p. 168.

stricted to the San Joaquin Valley, while the typical
C. wislizenii reaches the west slope of the Sierra Nevada
through Walker Pass, the summit of which is only 5,100?
feet in altitude and, therefore, not above the vertical
range of the species." It probably lives in the Sacramento Valley also, but has not been definitely recorded
from that region.

Genus 10. SAUROMALUS.

Sauromalus, DUMÉRIL, Arch. Mus. d'Hist. Nat., VIII, 1856, p. 535,
(type ater); *Euphryne*, BAIRD, Proc. Ac. Nat. Sci. Phila., 1858,
p. 253 (type obesus).

The head and body are much depressed, and but little
shorter than the heavy conical tail. All of the head-
plates are small. The labials are juxtaposed. The ear-
opening is large, with a very strong anterior denticulation. The dorsal scales are small and nearly uniform.
Long series of femoral pores and a strong transverse
gular fold are present. The lateral neck folds are
spinose. The superciliaries are juxtaposed.

13.—Sauromalus ater Duméril. ALDERMAN LIZARD. CHUCK-WALLA.

Sauromalus ater, DUMÉRIL, Arch. Mus. d'Hist. Nat., VIII, 1856, p.
536, pl. XXIII, figs. 3, 3a (no locality).
Euphryne obesus, BAIRD, Proc. Acad. Nat. Sci. Phila., 1858, p. 253
(type locality Fort Yuma, Cal.); BAIRD, U. S. Mex. Bound.
Surv., Rept. 1859, pl. XXVII.

Description.—Head and body very large, much depressed, the latter very broad. Head almost triangular,
with narrow rounded snout, and covered with small
plates largest on snout and temporal regions. Nostrils
opening upward, outward, and slightly backward, in
round plates a little nearer to end of snout than to
orbits. Superciliaries, like supraoculars, small and
juxtaposed. Suboculars all short, but strongly

keeled. Rostral plate very small. Labial plates small and of about equal size. Symphyseal plate long but very narrow. Several series of slightly enlarged sublabials passing gradually into the finely granular gulars. Gular fold covered with very small scales. Ear-opening large, almost vertical, with strong anterior denticulation of spinose scales. A strong fold on each side of neck, bearing numerous spinose tubercles. Scales on back and sides small, largest medially and on strong lateral fold, smooth and juxtaposed except laterally, becoming there keeled and slightly imbricate. Ventral scales smooth, smaller than dorsals. Tail little longer than head and body, conical, very stout, and covered with whorls of small, weakly keeled, feebly spinose scales. Femoral pores very large, varying in number from fifteen to eighteen.

The head, neck, and limbs are dull brownish black with a few scattered scales of grayish yellow. The back is dark brown or dull straw-color speckled with red, straw-color, or dark brown, and sometimes crossed by several broad bands of dark brown or black. The tail is dull straw-color with or without wide rings of black or dark brown. The ventral surfaces are black or dark brown more or less relieved with dull yellow.

A living specimen was colored as follows: The head and neck are uniform black, as are also the upper surfaces of the arms and legs. The hands and feet are speckled with dull yellowish white. The central portion of the back is chiefly brick-red dotted with black and yellowish white. Its lateral portions are chiefly black, but are dotted with deep vermilion and yellowish white. The sides are similar to the central portion of the back, but with less white and with red of a darker shade. The chest is black with a continuation of the red of each side crossing it and meeting its fellow on the median line just behind the insertion of the fore limbs. The belly is black, dotted and spotted with red. The lower surfaces of the limbs are black spotted with yellowish white and sparsely speckled with red. The tail is either all white or white crossed by wide bands of black.*

Length to anus	180	210
Length of tail	198	215
Snout to orbit	14	19
Snout to ear	36	45
Width of head	35	44
Fore limb	76	87
Hind limb	102	121
Base of fifth to end of fourth toe	32	38

Distribution.—The Chuck-walla or Alderman Lizard inhabits suitable situations throughout the Colorado and Mojave Deserts. It has been taken at Fort Yuma and on the eastern slope of the Julian Mountains in San Diego County, and thence occurs north to the desert ranges in the vicinity of Panamint and Death Valleys, California. It ranges east across southern Nevada (Pahrump Valley) to southwestern Utah.

Habits.—This lizard, the largest native to California, shares with several others the curious habit of defending itself with its tail. As this organ is very large and muscular, the animal can strike very quick and well-aimed blows, and does so with great vigor when teased. "It was generally found on lava or other dark rocks with which its coloration harmonized. It is a vegetarian, feeding entirely, so far as our observations go, on the

---

*There is much variation in the coloration of this lizard, especially as regards the black bands of the tail. These may be present or absent in the same individual at different times, and the change seems to be, at least to some extent, directly under the control of the animal. When the specimen whose colors are described above was put in a jar with chloroform, the black bands of the tail disappeared and reappeared several times before the lizard's death. Dr. Stejneger has observed the same color changes and thought them dependent upon the intensity of the light to which the animal is exposed.

buds and flowers of plants, with the addition sometimes of a few leaves. It is much prized by the Panamint Indians as an article of food. A number were eaten by members of our expedition, and their flesh was reported to be tender and palatable."[*]

Genus 11. UTA.

Uta, BAIRD & GIRARD, Proc. Ac. Nat. Sci., Phila., VI, 1852, p. 69 (type stansburiana). *Uro-saurus*, HALLOW., Proc. Ac. Nat. Sci. Phila., VII, 1854, p. 92 (type graciosus). *Phymatolepis*, DUMÉRIL, Arch. Mus. d'Hist. Nat., VIII, 1856, p. 548 (type bicarinatus).

The head and body are moderately depressed and much shorter than the tail. The head-plates are large, the largest (interparietal) being larger than the ear-opening. The dorsal scales are small and may be either uniform or heterogeneous. The labials are not imbricate. The ear-opening is large, with a strong anterior denticulation. One or more transverse gular folds and long series of femoral pores are present. The supercil-iaries are imbricate.

SYNOPSIS OF SPECIES.

a.—Dorsal scales or granules nearly equal, becoming gradually smaller laterally; an intense black band across shoulders or a blue patch behind axilla.

   b.—Dorsals juxtaposed, smooth; a narrow black band from shoulder to shoulder, above. .... **U. mearnsi.**—p. 64.

   b[2].—Dorsals imbricate, keeled; no black band from shoulder to shoulder; a round blue patch behind axilla. ..... **U. stansburiana.**—p. 66.

a[2].—Six or eight medial longitudinal series of dorsals largest, imbricate, keeled, becoming suddenly smaller laterally; no intense black band across shoulders or blue patch behind axilla.

   c.—Slender, tail more than twice head and body; enlarged dorsals sub-equal. ....**U. graciosa.**—p. 69.

   c[2].—Stouter, tail less than twice head and body; largest dorsals in four series, two on each side of the smaller median ones.

                                 U. symmetrica.—p. 70.

---

[*]Merriam, N. A. Fauna, No. 7, 1893, p. 174.

14.—Uta mearnsi Stejneger.   MEARNS'S LIZARD.

Uta mearnsi, STEJN., Proc. U. S. Nat. Mus., XVII, 1894, p. 589
(type locality Summit of Coast Range, United States and
Mexican Boundary line, California).

Description.—Head considerably depressed, snout very
low.   Canthus rostralis well marked, large nostrils
opening almost upward in rounded plates much nearer
to end of snout than to orbit.   Plates on head large,
smooth, and but slightly convex; interparietal largest.
Frontal plate usually divided transversely.   Two or
three posterior series of supraoculars enlarged, separated from frontals by one or two series of granules.
Superciliaries long and imbricate.   A long, narrow,
strongly keeled subocular, followed and preceded by
similar but smaller plates.   Rostral very wide and low,
as also the five or six supralabials.   Symphyseal plate
large and followed by several large chin-shields.   First
infralabials much larger than others.   Sublabials long
and narrow.   Skin on gular region covered with small,
smooth, rounded granules, slightly largest centrally and
near edge of strong gular fold.   A dermal fold on each
side between limbs.   Back and sides covered with
smooth, convex, rounded granules, largest medially,
smallest laterally, and changing gradually to small,
smooth, imbricate scales on belly.   Tail and anterior
and upper surfaces of limbs bearing larger imbricate
scales, each provided with strong keel ending in projecting spine.   Nineteen to twenty-five pores forming a
series along each thigh.   Males with enlarged postanal
plates.

The color above is bluish gray or olive, strongly
tinged with brown on the head and tail and crossed by
irregularly undulate bands of dark gray or slate.   A
narrow straight band of intense black crosses from

shoulder to shoulder over the back. The spaces between these bands are variously spotted, marbled, and reticulated with lighter. The limbs are irregularly

cross-banded with dusky. The basal two-thirds of the tail is pale brownish olive with wide blackish crossbars; its terminal third is uniform blackish. The lower surfaces are greenish white, suffused with bluish on the flanks, and reticulated with bluish gray or slate on the chin and throat.

Length to anus	75	79	82
Length of tail	140	150	
Snout to orbit			7
Snout to ear			20
Width of head			16
Fore limb		37	38
Hind limb		61	63
Base of fifth to end of fourth toe			22

Distribution.—This lizard has been found only on the eastern slope of the Coast Range of San Diego County California, near the Mexican boundary line. Here it is said to be " extremely plentiful " among rocks from the base to the summit of the range.

15.—Uta stansburiana Baird & Girard. BROWN-
SHOULDERED LIZARD.

Uta stansburiana, B. & G., Proc. Acad. Nat. Sci. Phila., VI, 1852,
p. 69 (type locality **Valley of Great Salt Lake, Utah**), *and*
Stansbury's Exped. Gt. Salt Lake, 1853, p. 345; pl. V, figs. 4-6.
Uta elegans, YARROW, Proc. U. S. Nat. Mus., 1882, p. 442 (type
locality **La Paz, L. C., Mex.**).

Description.—Body and head considerably depressed.
Snout low, rounded and rather short, with well
developed canthus. Nostrils large, opening upward
and outward, nearer to end of snout than to orbit.
Plates on head large, smooth, and usually more or less
convex, interparietal largest. Frontal plate usually
divided transversely. Three to five supraoculars enlarged and separated from frontals by one row of granules. Superciliaries long, somewhat projecting laterally,
and strongly imbricate. Central subocular very long,
narrow, and strongly keeled. Rostral and supralabials
very long and low. Other plates of upper surface of
head very irregular in size and position. Symphyseal
plate rather small, followed by three or four pair of
larger smooth plates separated from small infralabials
by from one to three series of moderately enlarged sublabials. Gular region covered with small, smooth, hexagonal scales, which change gradually into granules on
sides of neck, and into larger scales on strong transverse gular fold, where they are about the size of those
on belly. Edge of gular fold with a series of larger
projecting scales. Ear-opening with a strong denticulation of three or four pointed scales. Several longitudinal dermal folds usually present on sides of body
and neck. Back covered with
keeled scales of nearly uni-
form size, becoming gradually
granular on neck and sides. Largest scales on tail,

strongly keeled, sharply pointed, and larger above than below. Posterior surfaces of thighs and arms covered with small granular scales similar to those of sides of body. Other surfaces of limbs provided with large scales, keeled except on ventral surfaces of thighs, legs, and arms. Femoral pores varying in number from twelve to seventeen. About twenty-three to twenty-eight dorsal scales equaling the shielded part of head.

This lizard displays a very great amount of variation in both the pattern and intensity of its coloring. The back and sides are variously striped, spotted, or marbled with dark brown, blue, green, gray, or yellow; the former often with a double series of large brown spots, light-edged behind, which usually are much more distinct in females and young than in adult males. The tail is similarly marked, but is often ringed with brown. Below, the general color is yellowish white, usually more or less tinged with greenish or bluish on the sides of the belly. The throat in adults is blue, dotted or narrowly banded on the chin and sides with white, yellow, or orange. There is a round indigo spot behind the axilla and usually a brown patch in front of the shoulder.

Length to anus....................30	37	46	48	49	53
Length of tail..... 57	72		91	98	95
Snout to orbit...................... 3	3	3	4	4	4
Snout to ear 8	10	11	11	12	13
Width of head...................... 6	8	9	9	10	11
Fore limb.........................13	18	18	20	21	21
Hind limb.........................22	30	32	36	40	38
Base of fifth to end of fourth toe...... 9	13	13	15	16	15

Distribution.—The Brown-shouldered Uta is more generally distributed in southern California than any other lizard. It ranges over almost all of the desert and southern coast regions, besides the southern portion of the San Joaquin Valley and the lower levels of both slopes of the Sierra Nevada. It has been secured in San Diego (at Fort Yuma, Campo, San Diego, Witch Creek, Julian Mountains), Riverside (Indio, Cabazon, Banning, Riverside, San Jacinto, Hemet Valley), San Bernardino (Colton, Hesperia, Victor, Barstow, Needles, Leach Point Spring, Borax Flat), Los Angeles (San Pedro, Pasadena, Antelope Valley), Ventura (San Buenaventura), Santa Barbara (Santa Barbara), Kern (Mojave, Fort Tejon, Walker Pass, Kern River, Bakersfield, Caliente), Inyo (Lone Pine, Keeler, Coso, Panamint Mountains, Death Valley, Resting Spring, Round Valley), Fresno (Fresno, Pleasant Valley 10 miles west of Huron), Merced (five miles north of Los Banos), San Joaquin (five miles south of Lathrop), and San Benito (Bear Valley) Counties. It has also been found on Santa Cruz and San Clemente Islands.

It crosses Nevada (Dome Cañon, Virginia City, Pahrump Valley, Vegas Valley, Pahranagat Valley, Ash Meadows, Charleston Mts., Virgin River) to Utah (St. George, Valley Great Salt Lake), Oregon (Warner's Valley, Summer Lake), and Idaho (Snake River).

Habits.—The Brown-shouldered Lizard is a ground-loving species usually found in open fields or deserts or among rocks. Upon the approach of an enemy it quickly retires to some hole or crevice and shyly peeps out from time to time to see if the intruder has departed. At the old mission of Santa Barbara these graceful little lizards are especially tame and abundant, and live among the stones of the walls and fountains, darting in and out of the crevices which once were filled with mortar, sunning themselves on the sheltered surfaces, or chasing one another with all the abandon and apparent delight of children playing tag.

16.—**Uta graciosa** (Hallowell). Long-tailed Uta.

Urosaurus graciosus Hallow., Proc. Acad. Nat. Sci. Phila., 1854,
p. 92 (type locality "Lower [=Southern] California);" *id.*,
Rept. U. S. Pac. R. R. Surv., Vol. X, pt. IV, 1859, p. 4, pl. VII,
figs. 1a-1e.

Description.—Body and tail very slender, former as
well as head slightly depressed. Snout rounded but
rather narrow, with nostrils opening in small round
plates, much nearer to end of snout than to orbit.
Plates on head moderately large, smooth and almost flat;
interparietal largest. Frontal plate usually divided
transversely. Inner series of enlarged supraoculars
separated from frontal, frontoparietal, and parietal
plates by one or two rows of granules. Superciliaries
long, slightly projecting laterally, strongly imbricate.
Central subocular very long, narrow and strongly keeled.
Rostral and six or seven supralabials long and low.
Symphyseal plate moderately large and followed by a
series of plates separated from infralabials, except first
pair, by one or two series of sublabial plates. Gular
region covered with small, smooth, subhexagonal granules, which increase in size on, and are largest at edge
of, strong, transverse gular fold. Ear-opening large,
with denticulation of two or three scales, one being
much larger than the others. Sides of neck and body
more or less folded. About
five to eight rows of imbricate, keeled, equal-sized
scales forming a band down
middle of back and changing very abruptly to granules.
on its sides. Some scales on upper lateral fold enlarged.
Largest scales on the tail, larger above than below, and
strongly keeled but not pointed. Posterior surfaces of
thighs and arms covered with small granules similar to
those on sides of body. Superior and anterior surfaces

of limbs provided with keeled scales. Femoral pores about ten to twelve in number on each thigh. Tail more than twice as long as head and body.

The general color above is grayish, becoming darker on the sides and slightly tinged with yellow on the snout. On the back are rather indistinct, wide, blackish cross-bars, which are often interrupted on the vertebral line and sometimes alternate. The tail is grayish with faint narrow rings of brown or slate. The limbs are cross-barred with dusky above. The lower surfaces are silvery white more or less flecked with black or slate. Males have a yellow patch on the throat and a long blue area on each side of the belly.

Length to anus	33
Length of tail	75
Snout to orbit	3
Snout to ear	8
Width of head	6
Fore limb	12
Hind limb	22
Base of fifth to end of fourth toe	9

Distribution.—This lizard has been found only along the Colorado River. At Fort Yuma and The Needles, California, it is very rare and lives on or near the willows and mesquites, never on the open desert. In Nevada, it has been taken at Bunkerville, and at Callville, Lincoln County.

17.—Uta symmetrica Baird. TREE UTA.

Uta symmetrica, BAIRD, Proc. Acad. Nat. Sci. Phila., 1858, p. 253 (type locality **Fort Yuma, Cal.**).
? *Uta schottii*, BAIRD, l. c., p. 253 (type locality **Sta. Madelina, Cal.**).

Description.—Head and body considerably depressed. Snout rounded but rather narrow, with well developed canthi above which the nostrils open much nearer to end of snout than to orbits. Plates on head moderately large, smooth, and almost flat; interparietal largest. Frontal plate usually divided transversely. Inner series of enlarged supraoculars separated from frontal, fronto-parietal, and parietal plates by one or two rows of granules. Superciliaries long, very slightly projecting laterally, and strongly imbricate. Middle subocular very long, narrow, and strongly keeled. Rostral very wide and not very low. Four to seven long, low supralabials. Symphyseal plate moderately large and followed by series of large plates in contact with first pair of lower labials but separated from the others by one or two rows of sublabials. Chin and gular region covered with smooth subhexagonal granules, largest centrally and becoming imbricate on the strong transverse fold. Edge of fold with a series of projecting scales. Ear-opening large, with an anterior denticulation of from two to five pointed scales of much variation in size and shape. Two rows of medium-sized scales along middle of back, 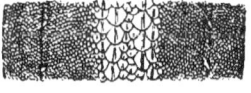 bordered on each side by two rows of much larger scales. Other dorsal scales very small, except a row of· widely separated enlarged scales on upper of two lateral dermal folds. Tail bearing whorls of strongly keeled and sharply pointed scales much broader above than below. Posterior surface of thighs and arms covered with small granules similar to those on sides of body. Superior and anterior surfaces of limbs provided with large keeled scales. Ventral scales smooth and about size of those on edge of gular fold. Femoral pores varying from ten to fourteen in number on each thigh. Thirteen to seventeen of largest dorsal scales equaling length of shielded part of head. Tail less than twice as long as head and body. Males with enlarged postanal plates.

The general color above is grayish or yellowish brown, paler and somewhat ochraceous on the head and the base of the tail, darkest along the upper lateral fold, and crossed by from six to eight light-edged bars of black or brown. These cross-bars are often very indistinct, usually interrupted on the middle of the back, and sometimes alternating with those on the opposite side. The light edgings of the dorsal bars may be either blue or yellow. The sides are often dotted with one or both of these colors. Narrow dark lines cross the top of the head, the most distinct being on the supraocular and frontal regions. In young specimens the dark coloring of the upper lateral fold is continued forward as a line, passing just above the ear-opening, crossing the orbit, and ending at the nostril. The tail is indistinctly ringed with dusky and often tinged with ochraceous. The lower surfaces are white, more or less dotted or suffused with dark brown or black. Males usually have a blue patch on each side of the belly and an area of lemon yellow, which sometimes acquires a tinge of blue, on the center of the throat.

Length to anus	30	36	49	54	55	59
Length of tail	49	58	85		95	102
Gular fold to anus	18	23	32	35	37	37
Snout to ear	8	9	11	12	12	13
Width of head	6	7	9	9	10	10
Fore limb	14	16	22	23	26	27
Hind limb	18	24	31	36	37	38
Base of fifth to end of fourth toe	7	10	14	14	14	16

Distribution.—The Tree Uta has been found in California only on the banks of the Colorado River near Fort Yuma, San Diego County. Its range seems not to extend much farther north, for a careful search failed to reveal its presence at The Needles in San Bernardino County.

Habits.—At Yuma this lizard is very abundant, but is rarely seen on the ground, preferring to climb over the rough bark of the willows or to hide between the planks of the railroad bridges. It feeds chiefly upon small insects.

Genus 12. SCELOPORUS.

Sceloporus, WIEGM., Isis, 1828, p. 369. " *Tropidolepis*, Cuv., R. A. (2), II, p. 38."

The head and body are slightly depressed and shorter than the tail. The head-plates are of moderate size, excepting the interparietal, which is very large. The dorsal scales are large, nearly equal sized, mucronate, and strongly imbricate. The ear-opening is large, with a well developed anterior denticulation. The labials are juxtaposed. There is no complete transverse gular fold, but a pouch is present on each side of the neck. Femoral pores are numerous. The superciliaries are imbricate.

SYNOPSIS OF SPECIES.

a.—Parietal and frontoparietal plates separated from enlarged supraoculars by a series of small scales or granules; scales on back of thigh smaller than those in front of anus.

  b.—Dorsal scales very small, forty-five to sixty-six on the middle of the back between interparietal plate and base of tail; scales on back of thigh smooth............................. ....S. graciosus.—p. 74.

  b².—Dorsal scales larger, thirty-five to forty-six on a line between interparietal and base of tail; scales on back of thigh keeled.*

   c.—Males with a blue patch on each side of the throat†; usually smaller.. ....S. occidentalis.—p. 77.

---

* Sometimes smooth in young.

† I have examined many hundreds of specimens of *S. occidentalis* and *S. biseriatus* and have not found a male of the latter with two blue throat-patches. Highly colored males of *S. occidentalis* are sometimes found in which the two blue patches have extended to, and even merged on, the median line, but by securing very young, or less brilliantly colored, males there should be no difficulty in determining which species occurs in a given locality, for such males never have a single median blue patch if they belong with *S. occidentalis*, and never have two lateral patches if referable to *S. biseriatus*. Females of the latter species have either one or two blue patches, while those of the more northern form usually have two or none.

c².—Males with one blue patch on center of throat; usually larger.

S. biseriatus.—p. 80.

a².—Parietal and frontoparietal plates in contact with enlarged supraoculars; scales on back of thigh not smaller than those in front of auus.

d.—A strongly contrasted black blotch or collar in front of shoulder; dorsal scales distinctly keeled, with long points.

S. magister.—p. 84.

d².—No distinct black blotch or collar in front of shoulder; dorsal scales less distinctly keeled, or smooth, with short points.

S. orcutti.—p. 86.

18.—Sceloporus graciosus Baird & Girard. MOUNTAIN LIZARD.

Sceloporus graciosus, B. & G., Proc. Acad. Nat. Sci. Phila., VI, 1852, p. 69 (type locality **Valley of Great Salt Lake, Utah**); *and* Stansbury's Rept. Exped. Gt. Salt Lake, 1853, p. 346, pl. V, figs. 1-3.

Sceloporus gracilis, BAIRD & GIRARD, Proc. Acad. Nat. Sci. Phila., VI, 1852, p. 175 (type locality **Oregon**); *and* GIRARD, U. S. Explor. Exped., Herp., p. 386, pl. 20, figs. 1-9.

Sceloporus vandenburgianus, COPE, Am. Nat., XXX, 358, Oct. 1896, p. 834 (type locality **Summit of Coast Range, San Diego Co., Cal.**).

Description.—Head and body somewhat flattened. Nostrils opening much nearer to end of snout than to orbits. Upper head-shields smooth, moderately large, and slightly convex; interparietal largest. Frontal usually divided transversely. Parietal, frontoparietal, and frontal plates separated from enlarged supraoculars by a series of small plates or granules. Superciliaries long, wide, and strongly imbricate. Middle subocular very long, narrow, and strongly keeled. Rostral plate very wide and rather high. Labials long, low, and almost rectangular. Below lower labials, some series of large sublabial plates. Symphyseal large and pentangular. Gulars small, smooth, imbricate, frequently emarginate posteriorly, about size of ventrals. Ear-opening large, slightly oblique, with an anterior denticulation of from four to seven accuminate scales. Dorsal scales equal-sized, keeled, pointed, larger than ventrals, and arranged in nearly parallel longitudinal rows. Scales on sides similar to those of back, but directed obliquely upward. No longitudinal dermal folds and no transverse fold on throat. Superior surfaces of limbs provided with keeled scales. Posterior surface of thigh covered with small, smooth scales. Ventrals smooth, but usually bicuspid. Caudal scales very much larger than dorsals, keeled and strongly pointed. Femoral pores varying in number from twelve to twenty on each thigh. Eleven to seventeen dorsal scales equaling length of shielded part of head. Number of scales in a row from interparietal plate to a line connecting posterior surfaces of thighs varying from forty-five to sixty-six; average in seventy-five specimens, fifty-five and one-half. Males with enlarged postanal plates.

The general color above is brown, olive, bluish or greenish gray, with one dorsal and two lateral series of closely set brown spots on each side. These spots have dark posterior and lateral edges, are usually larger and more distinct in females and young than in adult males, and are often more or less confluent, forming longitudinal bands separated by narrower bands of the lighter ground color. The head has no definite cross-lines, but the upper lateral band or series of spots is continued along the temple. The tail is very differently marked in different specimens, but usually shows traces of light and dark rings. Males have a large blue blotch, sometimes bordered internally with black, on each side of the belly, and the throat usually more or less washed with blue, which has a tendency to appear in narrow oblique lines. Females often lack the blue of the throat and sides of belly, but this color is sometimes present

and is not infrequently bordered above by a band of
bright reddish orange along each side of the body.

Length to anus..................... 23	44	48	50	56	63
Length of tail..................... 26	54	55	64	82	93
Snout to ear 6	10	11	11	12	12
Width of head.......... 6	9	10	10	11	11
Shielded part of head.. 5	9	10	10	11	12
Fore limb......................... 11	20	21	21	23	26
Hind limb............. 16	33	36	36	39	39
Base of fifth to end of fourth toe..... 7	14	14	15	15	16

Distribution.—This little lizard is a mountain-dwelling
species throughout its range in California, which extends the whole length of the State. It is very abundant in Hemet and Strawberry Valleys, in the San
Jacinto Mountains of Riverside County, but has not
been reported from any of the more northern coast
ranges.* On the western slope of the Sierra Nevada
the Mountain Lizard has been found from Kern River
north to Mount Shasta. It is common on the eastern
slope of these mountains opposite Mono and Owen's
Lakes and has been secured also in the Panamint
Mountains. In the southern Sierra Nevada its vertical
range extends from about 5,500 to nearly 9,000 feet above
the level of the sea.

It has been taken in Nevada (Juniper Mts., Mt. Ma-
gruder), western Utah, southern Idaho (Blackfoot, Big
Lost River, Lemhi Agency, and along Snake River from
Pocatello to Weiser), Washington (Puget Sound, Cowlitz
County), and Oregon (Dalles, Summer Lake, Grant's,
Umatilla, etc.). In the north it is not restricted to the
mountains.

Habits.—Nothing is known of the habits of this lizard
except that it is a ground-loving species. The eggs, laid

---

*Since this was written I have seen a specimen shot in Berryessa Valley, Napa County.

in June and July, are about 7x13 mm., each inclosed in a tough, leathery, non-calcareous shell.

19.—Sceloporus occidentalis Baird & Girard. BLUE-BELLIED LIZARD.

Sceloporus occidentalis, BAIRD & GIRARD, Proc. Ac. Nat. Sci. Phila., VI, 1852, p. 175 (type locality **California, probably Oregon**); GIRARD, U. S. Explor. Exped., Herp., p. 383, pl. 19, figs. 8–14.
Sceloporus frontalis, B. & G., Proc. Acad. Nat. Sci. Phila., VI, 1852, p. 175 (type locality **Puget Sound**); GIR., U. S. Expl. Ex., Herp., p. 384, pl. 19, figs. 1–7.
' *Sceloporus bocourtii*, BOULENGER, Cat. Liz. Brit. Mus., II, 1885, p. 229 [part] (**Monterey, Cal., Mt. Whitney, Cal., Santa Cruz**).

Description.—Head and body little depressed. Nostril opening much nearer to end of snout than to orbit. Upper head-shields smooth, moderately large, and slightly convex; interparietal much largest. Frontal usually divided transversely. Parietal, frontoparietal, and frontal plates separated from enlarged supraoculars by a series of small plates or granules. Superciliaries long and strongly imbricate. Middle subocular very long, narrow, and strongly keeled. Rostral plate of moderate height, but great width. Labials long, low, and nearly rectangular. Below lower labials and behind large pentangular symphyseal, some series of plates larger than gulars. Latter smooth, imbricate, and usually emarginate posteriorly. Ear-opening large, slightly oblique, with an anterior denticulation of smooth, accuminate scales. Scales on back equal-sized, keeled, pointed, and arranged in nearly parallel longitudinal rows. Scales on sides similar to those on back, but much smaller and directed obliquely upward. No longitudinal dermal folds and no transverse fold on throat. Upper surfaces of limbs provided with large, keeled scales. Posterior surface of thigh covered with small, accuminate,

keeled scales.   Ventral scales much smaller than dorsals,
smooth, imbricate, and usually bicuspid.   Tail furnished
with irregular whorls of strongly keeled and pointed
scales, much larger and rougher above than below.
Femoral pores varying in number from thirteen to twenty
on each thigh.   Seven to twelve dorsal scales equaling
length of shielded part of head.   Number of scales in
a row from the interparietal plate to a line connecting
posterior surfaces of thighs varying from thirty-five to
forty-six; average in thirty specimens, forty-one and
eight-tenths.   Males with enlarged postanal plates.

The color above is grayish, brownish, or olive, usually
with one series of crescent-shaped or triangular brown
spots, edged posteriorly with pale blue or green, on each
side.   A paler longitudinal band usually separates the
dorsal and lateral regions.   The sides are brownish or
buffy, mottled with darker brown and dotted with green
or pale blue.   Narrow brown lines cross the head, but
are more or less interrupted.   A brown line connects
the orbit and upper corner of the ear and is continued
backward on the neck.   A large blue patch on each side
of the belly is usually bordered internally by a black
band of varying width.   The throat is white, more or
less dotted or suffused with slate or black, and with or
without a blue patch on each side.   In highly colored
males the black bands of the belly meet medially and
the throat is intensely black with large round blue
patches, which sometimes merge on the median line.
The chest is white or yellowish, often dotted or suffused
with black.   The preanal region and the lower surfaces
of the limbs are white, sometimes dotted or tinged with
slaty-black.   The posterior surfaces of the limbs are
yellowish, deepest on the thighs, along the back of
which runs a dark line.   In young, and some females,

the green edging of the dorsal spots is replaced by gray
or buff.

Length to anus.	27	44	67	68	68	70
Length of tail	31	64	94	88	104	93
Snout to ear	7	11	15	14	15	14
Width of head	6	9	13	13	13	13
Shielded part of head	7	10	14	14	14	14
Fore limb	12	20	30	28	30	30
Hind limb	17	33	47	44	48	48
Base of fifth to end of fourth toe	7	14	20	16	20	20

Distribution.—The Blue-bellied Lizard is a northern
species, which, coming to us from Oregon, occupies
the long coastwise strip of California lying to the west
of the Sacramento and San Joaquin Valleys. Close to
the coast, its range extends as far south as Ventura
County. In the north, its territory stretches eastward
at least as far as Mount Shasta and probably extends
some distance south in the Sacramento Valley and on
the western slope of the Sierra Nevada. Farther south,
it crosses the San Joaquin Valley to the western slope
of the Sierra Nevada. I have examined specimens from
Siskiyou (Sissons, Fort Jones), Mendocino (Fairbanks),
Lake (Kelseyville), Sonoma (Healdsburg, Santa Rosa),
Napa (Calistoga, Napa, Ætna Springs), Marin (San
Anselmo, Mill Valley), Contra Costa (Mount Diablo,
Crockett), Alameda (Oakland, Calaveras Valley, Liver-
more, Altemonte), San Francisco, San Mateo (Pesca-
dero, La Honda), Santa Clara (Palo Alto, Black Moun-
tain, Santa Clara, Los Gatos, Alum Rock Cañon, Smith
Creek, Canada Valley), Santa Cruz (Soquel), Monterey
(Monterey, Pacific Grove, Pleyto), San Benito (Bear
Valley, San Benito Valley), San Luis Obispo (San Mi-
guel, San Luis Obispo), San Joaquin (Tracy, San
Joaquin Bridge), Merced (near Merced), Tuolumne
(Big Oak Flat, Groveland to Crocker's, Hodgdon's),

Fresno (Los Gatos Cañon), Santa Barbara (Santa Barbara), and Ventura (San Buenaventura) Counties,
California.

It ranges north across western Oregon and Washington to the Straits of Juan de Fuca.

Habits.—The Blue-bellied or Western Fence Lizard is
by far the most numerous of its tribe in western central
California. It is usually to be found about fences, piles
of wood or stone, the great brush-heap homes of the
wood-rat *(Neotoma)*, or roadside banks honeycombed
with abandoned gopher *(Thomomys)* holes, which afford
it ample opportunity to hide upon the approach of
danger. Its coloration, especially the intensity of the
black of the lower surfaces and the blue of the throat,
is subject to much variation in the same individual, and
is more or less dependent upon the coloring of surrounding objects.

In winter it is sometimes found in the interior of decaying logs, but I believe that it more frequently hiber-
nates under ground.

20.—**Sceloporus biseriatus** Hallowell.   FENCE LIZARD.

> *Sceloporus bi-seriatus,* HALLOW., Proc. Acad. Nat. Sci. Phila , 1854,
> p. 93 (type locality border of El Paso Creek and in Tejon
> Valley [California]); *and* U. S. Pac. R. R. Surv., Rept., X,
> pt. IV, 1859, p. 6, pls. VI, figs. 2a–2f, & VIII.
>
> *Sceloporus longipes,* BAIRD, Proc. Acad. Nat. Sci. Phila., 1858, p. 254
> (type locality Fort Tejon, Cal.).
>
> *Sceloporus smaragdinus,* COPE, Wheeler's Surv. W. 100th Meridian,
> 'V, p. 572, pl. XXIV, figs. 2, 2a (type localities Beaver, Utah;
> Nevada; Dome Canon, Utah).
>
> *Sceloporus bocourtii,* BOULENGER, Cat. Lizards Brit. Mus., II,
> 1885, p. 229 [part], type localities Monterey, Cal., Mt.
> Whitney, Cal., Santa Cruz).

Description.—Head and body little depressed. Nostril
opening much nearer to end of snout than to orbit.

Upper head-shields smooth, moderately large, and very slightly convex; interparietal much largest. Frontal usually divided transversely. Parietal, frontoparietal, and frontal plates separated from enlarged supraoculars by a series of smooth plates or granules. Superciliaries long, wide, and strongly imbricate. Middle subocular very long, narrow, and strongly keeled. Rostral plate of moderate height, but great width. Labials long and low. Below lower labials and behind large pentangular symphyseal some series of plates larger than gulars. Latter of moderate size, smooth, imbricate, and usually emarginate posteriorly. Ear-opening large, slightly oblique, with an anterior denticulation of smooth accuminate scales. Back with equal-sized, keeled, pointed scales arranged in nearly parallel longitudinal rows. Scales on sides similar to those on back but much smaller and directed obliquely upward. No longitudinal folds and no transverse fold on throat. Superior surfaces of limbs provided with large, keeled scales. Posterior surface of thigh covered with small, accuminate, keeled scales. Ventral scales much smaller than dorsals,

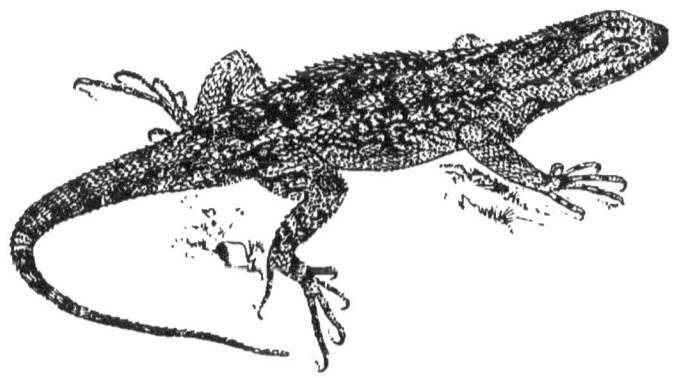

smooth, imbricate, and usually bicuspid. Tail with irregular whorls of strongly keeled and pointed scales much larger and rougher above than below. Femoral pores varying in number from thirteen to eighteen on each thigh. Seven to eleven dorsal scales equaling length of shielded part of head. Number of scales in a row from interparietal plate to a line connecting posterior surfaces of thighs varying from thirty-five to forty-four; average in thirty specimens, forty and two-tenths. Males with enlarged postanal plates.

The back is brown, olive, or grayish buff, marked with large blotches or undulate cross-bands of dark brown and more or less dotted, spotted, or blotched with green or pale blue. The sides are similarly colored. Above, the head is brown or olive with narrow lines of dark brown, which are most distinct between the eyes and on the temples. The tail is olive or brown with irregular dark brown rings. All the lower surfaces are grayish or yellowish white, often suffused with slate or dull black. Along each side of the belly is a large patch of deep blue, usually bordered internally by a black band of varying width. Males have one large central throat-patch of deep blue, but females may have two lateral patches. The posterior surfaces of the limbs are yellow.

Length to anus	41	45	68	72	77	78
Length of tail	64	69	108	103		
Snout to ear	10	11	15	15	17	17
Width of head	8	9	13	13	14	14
Shielded part of head	9	11	14	14	15	15
Fore limb	19	23	32	31	35	35
Hind limb	31	35	51	51	56	55
Base of fifth to end of fourth toe	13	15	22	20	21	22

Distribution.—The Fence Lizard, or Tree Swift as it is sometimes called, occupies the coast region south of

the range of *Sceloporus occidentalis* and both slopes of
the Sierra Nevada, together with portions of the San
Joaquin Valley and the desert ranges farther east.
How far north it lives on the western slope of the
Sierra Nevada and where it meets *Sceloporus occidentalis*,
we do not know. It doubtless occurs throughout the
whole length of the Great Basin, for it is common in
Idaho. Its vertical range in central California extends
up to about 8,000 feet.

I have examined specimens from San Diego (San
Diego, Mexican border between Campo and the coast,
Cuyamaca Mts., Witch Creek, Santa Ysabel Valley
3,000–4,000 feet, Julian Mts.), Riverside (Cahuilla
Valley, Strawberry Valley, Hemet Valley 5,000 feet, San
Jacinto, Temescal, Riverside), Los Angeles (Alhambra,
Pasadena), San Bernardino (Mojave River near Victor,
Warren's Wells, Lytle Creek), Kern (Fort Tejon, Tejon
Pass, Walker Basin, Havilah, Kernville, Walker Pass),
Tulare (South Fork Kern River, Tulare, Visalia, Three
Rivers, East Fork Kaweah River 1,650–5,200 feet, Shot-
gun Cañon, Kern River Lakes 7,000 feet), Fresno
(Fresno, Horse Corral Meadows, San Joaquin River
7,500), Mariposa (Nevada Falls Yosemite Valley, Mari-
posa), and Inyo (Coso Mts., Argus Mts., near Owen's
Lake, Mt. Whitney, Independence Creek, White Mts.,
Round Valley) Counties, California.

It crosses Nevada (Charleston Mts., Mt. Magruder,
Juniper Mts., Grapevine Mts., Pyramid Lake) to Utah
and Idaho (along Snake River), and probably occurs in
eastern Oregon.

Habits.—Like its northern congener—*S. occidentalis*—
and its larger relative of the desert—*S. magister*—the
Fence Lizard frequently performs a curious exercise
while watching an intruder and determining whether

he be friend or foe. Clinging to the rough bark of a
tree or the lichen-painted surface of some old fence, it
rapidly raises and lowers its head and body, often at-
tracting attention to itself where the harmony of coloring would prevent its being noticed if motionless. It
is rarely seen in open fields, preferring wooded districts
or areas where rocks abound.

21.—Sceloporus magister Hallowell. SCALY LIZARD.

Sceloporus magister, [*] HALLOW., Proc. Acad. Nat. Sci. Phila., 1854,
    p. 93 (type locality near Fort Yuma, California); *and* U. S.
    Pac. R. R. Surv. Rept., X, pt. IV, 1859, p. 5; STEJNEGER, N. A.
    Fauna, No. 7, 1893, p. 178, pl. I, figs. 2a–2c.

Description.—Head and body little depressed. Nasal
opening slightly nearer to end of snout than to orbit.
Upper head-plates smooth, often a little convex, and
usually slightly imbricate; interparietal largest. Frontal
divided transversely. Parietal and (usually) frontoparietal plates not separated from large supraoculars.
Latter very broad, as also the strongly imbricate
superciliaries. Middle subocular very long, narrow, and
strongly keeled. Rostral plate wider than high. Labials
long but very low, inferior larger than superior. Symphyseal large, followed by several plates larger than
gulars and separated from lower labials by from one to
three rows of narrow sublabials. Gular region with
scales smooth, flat, bicuspid, and strongly imbricate, as
also belly. Ear-opening large, nearly vertical, and protected by a series of very long accuminate scales. Back
with equal-sized, rather weakly keeled, but strongly
pointed, scales arranged in nearly parallel longitudinal
rows. Scales of sides pointed obliquely upward and
changing gradually from carinate dorsals to smaller
smooth ventrals. No longitudinal dermal folds. Upper

---

*This species was long confused with *S. clarkii,* which is not Californian.

surfaces of limbs provided with strongly keeled and pointed scales. Scales on posterior surface of thigh large, accuminate, strongly keeled and pointed. Upper caudal scales similar to dorsals, but having longer points. Femoral pores varying in number from eleven to fifteen on each thigh. Five to ten dorsal scales equaling length of shielded part of head. Number of scales in a row from interparietal plate to a line connecting posterior surfaces of thighs varying from twenty-nine to thirty-five; average in thirty specimens, thirty-one and two tenths. Males with enlarged postanal plates.

The back is gray, yellow, brown, or copper-color, without distinct markings or with a very broad (4–5 scales) band of dark brown along its anterior half in adult males, irregularly spotted or blotched with dark brown in females and young. There is a strongly contrasted black bar or collar in front of each shoulder. Faint indications of dark lines may sometimes be seen on the head. The tail is brown or olive with indistinct brown rings or cross-bars. In highly colored males, the throat has a central patch of blue, which gradually fades anteriorly and changes to black posteriorly. The belly has lateral bands of deep blue, more or less bordered or replaced with black. The scales of the sides are variously tinted with black, blue, green, yellow, orange, red, and brown. There usually is a black area in front of the thigh. In females and young, the throat and belly are usually white or grayish yellow.

Length to anus.................... 35	50	68	88	106	109
Length of tail..................... 51	67	99	112	149	158
Snout to ear...................... 10	12	16	20	24	25
Width of head..................... 9	10	14	16	24	25
Shielded part of head............. 10	11	15	17	20	20
Fore limb......................... 17	24	31	42	45	53
Hind limb......................... 27	35	49	58	68	72
Base of fifth to end of fourth toe..... 11	14	19	23	25	28

Distribution.—The Scaly Lizard may be considered a member of the desert fauna, although I secured a specimen near the mouth of the Los Gatos Cañon, about six miles above Coalinga, in the southwestern part of Fresno County. It is not rare along the Colorado River, both at Fort Yuma in San Diego County and at The Needles in San Bernardino County. In the latter county, it has been taken also near the Mojave River at Barstow, at Victor, at Warren's Wells, among the tree-yuccas at Hesperia, and on the desert near the base of Granite Mountains. Specimens have been taken at Mojave Station and Walker Pass, in Kern County, and near Lone Pine and in the Panamint and Argus Mountains in Inyo County, California.

It crosses southern Nevada to southwestern Utah, the most northern locality at which it has been taken being the Big Bend of the Truckee River, Nevada.

Habits.—This large lizard is rarely seen on the open desert, preferring the shelter of yuccas, mesquites, cottonwoods, and willows, about which it climbs with great agility. Dr. Merriam says it "is a mixed feeder, both insects and flowers being found in the stomachs examined."

22. — **Sceloporus orcutti** Stejneger.   DUSKY SCALY LIZARD.

Sceloporus orcutti, STEJN., N. A. Fauna, No. 7, 1893, p. 181 (footnote) pl. 1, figs. 4a–4c (type locality **Milquatay Valley, San Diego County, California**).

Description.—Head and body much depressed. Nasal opening a little nearer to end of snout than to orbit. Upper head-plates smooth and usually somewhat convex, supraoculars often slightly imbricate. Frontal divided transversely. Parietal and frontoparietal plates not separated from large supraoculars. Latter very broad, as also the strongly imbricate superciliaries. Middle subocular very long, narrow, and strongly keeled. Rostral plate much broader than high. Labials long but very low, inferior slightly larger than superior. Symphyseal large and followed by several plates larger than gulars and separated from lower labials by from one to three rows of narrow sublabials. Gular region with scales smooth, flat, bi- or tricuspid, and strongly imbricate, as also belly. Ear-opening large, nearly vertical, and protected by a series of long, accuminate scales. Scales of back in nearly parallel longitudinal rows, equal-sized, with no keels or very obtuse ones, and points which scarcely protrude beyond the serrate posterior outline. Scales of sides pointed obliquely upward and changing gradually from smoother dorsals and smaller smooth ventrals, becoming keeled and strongly pointed. No longitudinal dermal folds. Upper surfaces of limbs provided with strongly keeled and pointed scales. Scales on posterior surface of thigh large, accuminate, strongly keeled and pointed. Upper and lateral caudal scales nearly smooth, but with very long points. Femoral pores varying in number from twelve to sixteen on each thigh. Six to twelve dorsal scales equaling length of shielded part of head. Number of scales in a row from the interparietal plate to a line connecting posterior surfaces of thighs, thirty-one to thirty-seven; average in fifteen specimens, thirty-three and six-tenths. Males with enlarged postanal plates.

In very young specimens the back is crossed by numerous dark brown bands separated by narrower ones of paler brown. The narrow bands gradually become more or less greenish or bluish and some of the dorsal scales become copper-color with blue centers. In adult males the cross-bands have almost or entirely disappeared and the back and sides are finely mottled with brown, gray, green, blue, and copper-color. The upper head-plates are brown with pale centers. The tail is cross-barred with dark and light brown or green. The throat and belly of young specimens are bluish or yellowish white with oblique dusky bands corresponding to those on the sides of the head and body. In adult males the throat and belly are nearly uniform dull purplish cyanine blue, the edges of the scales often being black or reddish brown. There is a slightly darker area in front of the shoulder, but no distinct blotch or collar is present.

Length to anus....................39	72	86	100	106	109
Length of tail.....................52	102	118	115	119+	122+
Snout to ear 10	17	17	20	20	21
Width of head 9	15	17	20	21	21
Shielded part of head..............10	15	16	18	19	19
Fore limb.........................19	34	.39	45	44	48
Hind limb........................28	52	56	66	64	67
Base of fifth to end of fourth toe 11	20	22	24	23	25

Distribution.—The Dusky Scaly Lizard has been found only in the coast ranges of Riverside and San Diego Counties, California, and in the northern part of Lower California. Originally described from Milquatay Valley, it has since been secured between Campo and the coast, at Witch Creek, in Clogston's Valley, Cahuilla Valley, Strawberry Valley, Hemet Valley, and at San Jacinto, Riverside, and Temescal.

Habits.—This lizard of the rocks is common near San Jacinto, but is very timid, rarely permitting the collector to approach near enough to use fine shot with deadly effect. In the cool of the morning and late in the afternoon it may be seen upon the highest point of some rounded boulder, but during the warmer hours it avoids the direct rays of the sun and must be sought on the shady sides of the granite, into whose crevices it quickly disappears when approached too closely.

Genus 13. PHRYNOSOMA.

Phrynosoma, WIEGM., Isis, 1828, p. 367; *Batrachosoma*, FITZINGER, Syst. Rept., 1843, p. 79 (type coronatum); *Anota*, HALLOW., Proc. Ac. Nat. Sci. Phila., 1852, p. 182 (type m'callii); *Doliosaurus*, GIRARD, U. S. Explor. Exped., Herp., 1858, p. 407.

The body is very broad, greatly depressed, without dorsal crest, but usually with a lateral fringe. The head is covered with small subequal scales and bears bony spines on the occipital and temporal regions. The tympanum is either distinct or partially or entirely scaled. The dorsal scales are very irregular in size and shape. Series of femoral pores and one or more transverse gular folds are present. The tail is short. Males have enlarged postanal plates.

SYNOPSIS OF SPECIES.

a.—Nostrils opening on or almost on the lines joining the supraorbital ridges with the end of the snout.

  b.—Gular scales small, nearly equal-sized; a series of enlarged scales below, but not much larger than, the lower labials; occipital spines very short....................................**P. douglassii.**—p. 90.

  b².—Several longitudinal series of enlarged pointed gular scales; a series of very large spinose plates below the lower labials; head-spines large.

    c.—Head-shields convex and almost smooth.

                P. blainvillii.—p. 91.

    c².—Head-shields flat, with numerous ridges and granulations.

                P. frontale.—p. 93.

a².—Nostrils opening well above the lines joining the supraorbital ridges with the end of the snout; a series of very large shields below the lower labials; gular scales small, equal or with one row of enlarged scales on each side.

  d.—One series of small peripheral spines; six to twelve femoral pores; no narrow dark median dorsal line....**P. platyrhinos.**—p. 98.

  d².—Two or three series of peripheral spines; eighteen to twenty femoral pores; a narrow dark median dorsal line.

              P. m'callii.—p. 100.

23.—Phrynosoma douglassii (Bell). PIGMY HORNED
TOAD.

Agama Douglassii, BELL, Trans. Linn. Soc., XVI, p. 105, pl. X (type
locality Columbia River).

Phrynosoma douglassii, GIRARD, Stansbury's Exped. Gt. Salt Lake,
1853, p. 362, pl. VII, figs. 6–10.

Phrynosoma douglassi pygmœa, YARROW, Proc. U. S. Nat. Mus., V.
1882, p. 443 (localities Ft. Walla Walla, Wash.; Des Chutes
River, Oreg.; Ft. Steilacoom): YARROW, Bull. U. S. Nat.
Mus., No. 24, 1882, p. 70 (type locality stated as Des Chutes R.,
Oreg.); TOWNSEND, Proc. U. S. Nat. Mus., X, 1887, p. 238.

Description.—Nostrils opening on lines joining superciliary ridges with end of snout. Gular scales small
and nearly equal-sized. A series of enlarged sublabial
scales not much larger than infralabials, separated from
latter by several rows of granules. Head-spines very
short; four or five temporals, one occipital, and one
postorbital on each side. Occipital spines nearly erect.
Supralabials small but prominent. Infralabials slightly
larger than supralabials and continued farther back,
becoming gradually spinose. Other head-scales small,
irregular in size and arrangement, more or less convex,
and roughened with ridges and granulations. Two
groups of spines on neck, upper being larger. Back,
tail, and upper surfaces of limbs with scattered, large,
more or less erect, keeled, tubercular scales; between
these, skin covered with smaller scales and granules.
Body with fringe of one series of peripheral spines.
Chest and belly and lower surfaces of hind limbs and tail
covered with small smooth scales. Tympanum either
naked or scaled. Long series of femoral pores almost
meet medially. Males sometimes with enlarged postanal
plates.

The back is olivaceous, yellow, brown, gray, or slate,
with two or four rows of dark blotches. These blotches
vary greatly in intensity but are almost always edged

posteriorly with white, gray, or yellow. There is an indistinct large dark blotch on each side of the neck. The
coloring of the tail is similar to that of the back. The
ground color of the head is very variable, as are also its
darker markings. The entire lower surface is white or
pale yellow, sometimes faintly marked with gray or
slate.

Length to anus...	24	46	58	60	60	61
Length of tail...	9	25	29	30	25	28
Snout to ear...	6	12	15	15	14	15
Width of head...	8	14	16	17	17	18
Length of occipital spine...	1	1⅓	1⅓	1⅓	1⅓
Fore limb...	12	20	25	24	24	24
Hind limb...	16	26	34	33	33	33
Base of fifth to end of fourth toe...	5	8	10	10	10	10

Distribution.—Washington, Oregon, and Idaho contain the greater part of the range of this lizard. Mr.
Chas. H. Townsend has recorded it from the northern
base of Mount Shasta, California, which is, I believe,
the only definite locality at which the Pigmy Horned
Toad has been found in this State, although it probably
occurs in many parts of Siskiyou and Modoc Counties.

It has been recorded from Fort Steilacoom, Fort Walla
Walla, and North Yakima, Washington; from Grant's,
Des Chutes River, Willamette Valley, and between
Warner's and Goose Lakes, Oregon; and from Blackfoot,
Big Butte, Big Lost River, Pocatello, vic. Lewiston,
Conant, Arco, and American Falls, Idaho.

24. — Phrynosoma blainvillii Gray. BLAINVILLE'S
HORNED TOAD.

Phrynosoma Blainvillii, GRAY, Zool. Beechey's Voy., 1839, p. 96,
pl. XXIX, figs. 1 (type locality California).*

Description.—Nostrils opening on lines joining superciliary ridges with end of snout. Head-spines large;

---

* Many authors have confused this species and *P. frontale* with *P. coronatum* Blain., of
Lower California, which does not live in California.

three to six temporals, one occipital, and one postorbital on each side, and one small interoccipital. Sometimes with small spines above and between temporals and often in front of occipital spines. Temporal scales with ridges running in the general direction of temporal spines. Other upper head-scales convex and almost or quite smooth. Several longitudinal series of gular scales enlarged and spinose, but becoming smaller toward median line, and continued on gular fold or folds. A series of five or six spinose sublabials, often continued posteriorly by smaller plates. Below corner of mouth, a very broad spine followed by a long slender one. Two groups of spines on each side of neck, lower usually larger. Back and tail with large, scattered, somewhat elevated, keeled, tubercular scales, between which smaller scales and granules. Two rows of peripheral spines; lower series shorter than upper and composed of smaller spines. Tail edged with a single row of lateral spines and bearing a small group of slender spines just behind thigh. Scales on anterior surfaces of limbs large, pointed, and strongly keeled. Those on chest, abdomen, and proximal part of ventral surface of tail smooth, but those on terminal portion of tail keeled. Tympanum not covered with scales. Long series of (12 to 18) femoral pores present. Males usually with enlarged postanal plates.

The ground color above is brownish, yellowish, reddish, or grayish, usually darker laterally. A large brown patch occupies each side of the neck. On the back are undulate cross-bands or large irregular spots of dark brown, usually edged posteriorly with yellow or white. Similar markings are seen on the tail. The head is usually yellow, but may be clouded with slate. Its larger spines are often reddish. The lower surfaces are yellow

or yellowish white, uniform or mottled with slate or
gray. All markings are usually more distinct in young
than in old specimens, but are very variable in both.

Length to anus 29	74	68	88	92	98
Length of tail.......................13	40	40	43	38	47
Snout to ear ....,........ 8	15	17	18	18	18
Width of head*....................11	26	30	30	30	32
Length of occipital spine............ 2	6	10	11	9	9
Fore limb...........................14	34	38	39	38	40
Hind limb.........................19	44	52	54	52	53
Base of fifth to end of fourth toe...... 6	14	15	17	15	15

Distribution.—Blainville's Horned Toad is an inhab-
itant of the coastal slopes of San Diego, Riverside, San
Bernardino, and Los Angeles Counties, California,
where it has been taken at San Diego, Twin Oaks, Oak.
Grove, Clogston's Valley, San Jacinto, Hemet Valley,
Banning, Riverside, Lytle Creek, Warren's Wells,
Ontario, Pasadena, and Alhambra. It has not been
collected on the desert proper and doubtless does not
live there, although it is not rare in San Gorgonio Pass.
Where it meets *P. frontale* is not known.

25.—Phrynosoma frontale Van Denburgh. CALIFOR-
NIA HORNED TOAD.

> *Phrynosoma frontalis,* VAN D., Proc. Calif. Acad. Sci., Ser. 2, IV,
> p. 296 (type locality **Bear Valley, San Benito County,
> California**).

Description.—Nostrils open on lines joining superciliary ridges with end of snout. Head-spines usually a
little smaller than those of *P. blainvillii;* three to six
temporals, one occipital, and one postorbital on each
side, and one small interoccipital. Sometimes with
small spines above and between temporals and usually
in front of occipitals. Temporal scales with ridges run-

---

*To tips of temporal spines.

ning in the general direction of temporal spines. Other upper head-scales flat, each with numerous ridges and granulations usually darker than ground color of head. Several longitudinal series of gular scales enlarged and spinose, but becoming smaller toward median line, and continued back onto gular fold or folds. A series of five or six large spinose sublabials, often continued posteriorly by smaller plates. Below corner of mouth, a very broad spine followed by a long slender one. Two groups of spines on each side of neck, lower usually larger. Back and tail bearing large, scattered, elevated, keeled, tubercular scales, between which smaller scales and granules. Two rows of peripheral spines; lower series shorter than upper and composed of much smaller spines. Tail bordered with a single row of lateral spines and having a small group of slender ones just behind thigh. Scales on anterior surfaces of limbs large, pointed, and strongly keeled. Those on chest, belly, and proximal part of the tail smooth, but those on terminal part of tail keeled. Tympanum not covered with scales. Long series of femoral pores present. Males usually with enlarged postanal plates.

The upper surfaces are variously tinted with yellow, brown, red, gray, or slate. A large brown patch occupies each side of the neck. On the back are undulate cross-bands or large blotches of dark brown, usually edged posteriorly with white or yellow. Similar markings may be seen on the tail. The keels of many of the large dorsal tubercles are dark brown. The head is yellowish, usually dotted with brown. Its larger spines are sometimes reddish. The lower surfaces are yellow or yellowish white, uniform or mottled with slate or gray. All markings are usually more distinct in young than in old specimens, but are very variable in both, the intensity of coloring depending very greatly upon the color of surrounding objects and changing in the same individual in the course of a very few minutes.

Length to anus......................	31	38	55	74	87	89
Length of tail.....................	10	15	22	34	40	36
Snout to ear......................	6	8	11	15	16	17
Width of head*.......	9	13	17	23	26	26
Length of occipital spines...........	2	3	5	7	6	6
Fore limb........................	13	17	24	34	33	36
Hind limb......................	17	22	32	47	46	51
Base of fifth to end of fourth toe.....	5	7	10	14	13	16

Distribution.—This lizard occupies a much larger area in California than does *Phrynosoma blainvillii*. The most southern localities from which I have seen specimens are in Kern County (Fort Tejon, Kernville, Bakersfield). Thence it ranges north across Tulare, Kings, Fresno, Merced, and San Joaquin Counties. Farther west, it has been taken in San Benito (Bear Valley), Monterey (Monterey, Pacific Grove), Santa Clara (Cañada Valley, Coyote Creek, Santa Clara, Mayfield, Los Gatos, Wright's, Morgan Hill), and San Mateo (Searsville) Counties. We do not know how much farther north this horned toad lives, but Dr. Boyle collected one in El Dorado County. It does not occur east of the Sierra Nevada.

Horned toads from Santa Barbara and Ventura Counties will probably be found to be intermediate between this and the preceding species, in which case this form must be known as *Phrynosoma blainvillii frontale*.

Habits.—Their grotesqueness of form, slowness of movement, and the ease with which they may be fed, cause these lizards to be much sought as pets. In confinement they are usually very docile and become so tame that they will readily take flies or other small

---

*To tips of temporal spines.

insects from the fingers of their keeper. Individuals
which have been recently caught, however, often show
considerable anger when handled, puffing themselves
up and hissing fiercely, seizing their tormentor's fingers
with their impotent jaws, or throwing at him a stream
of blood from the corner of the eye. It is said that the
Mexicans call them sacred toads because they weep tears
of blood. The best account of this most curious habit
has been given us by Dr. O. P. Hay,* who, writing of a
specimen of *Phrynosoma frontale*, says, in part:

"About the first of August it was shedding its outer
skin and the process appeared to be a difficult one, since
the skin was dried and adhered closely. One day it oc-
curred to me that it might facilitate matters if I should
give the animal a wetting; so, taking it up, I carried it
to a wash-basin of water near by and suddenly tossed
the lizard into the water. The first surprise was probably experienced by the *Phrynosoma*, but the next surprise was my own, for on one side of the basin there
suddenly appeared a number of spots of red fluid, which
resembled blood. * * * A microscope was soon pro-
cured and an examination was made, which immediately
showed that the matter ejected was really blood. * * *
There appeared to be a considerable quantity of the
blood, since on the sides of the vessel and on the wall
near it I counted ninety of the little splotches. * * *
The next day * * * I picked up the lizard and was
holding it between my thumb and middle finger, and
stroking its horns with my fore finger. All at once a
quantity of blood was thrown out against my fingers,
and a portion of it ran down the animal's neck; and
this blood came directly out of the right eye. It was

*Proc. U. S. Nat Mus., XV, 1892, p. 375.

shot backward and appeared to issue from the outer canthus. It was impossible to determine just how much there was of the blood, but it seemed that there must have been a quarter of a teaspoonful. I went so far as to taste a small quantity of it, but all I could detect was a slight musky flavor.

" Mr. Denton * * * has communicated to me his experience with the Horned Toad * * * at Sonora, Cal. * * * He was gently stroking the animal on the back, when it appeared to look at him as if taking aim, and then, all at once, a stream of blood was shot into his eye. There was so much of it that it ran down on his shirt bosom. He thought there was between a tablespoonful and a teaspoonful. The blood was shot out with so much force that some pain was produced, and there was pain felt for some little time, though this ceased as soon as the blood was wiped out. The next morning the eye was somewhat inflamed, but this condition soon passed away. Not long afterwards, perhaps the next morning, the animal squirted blood out of the other eye."

Mr. Vernon Bailey, who caught the horned toad which afterwards became the subject of Dr. Hay's article, writes:*

" On taking it in my hand a little jet of blood spurted from one eye a distance of 15 inches and spattered on my shoulder. Turning it over to examine the eye another stream spurted from the other eye. This he did four or five times from both eyes until my hands, clothes, and gun were sprinkled over with fine drops of bright red blood. * * * About four hours later * * * it spurted three more streams from its eyes."

---

*N. A. Fauna, No. 7, 1893, p. 189.

I myself have observed this strange performance twice, only in these instances the blood was not projected forcibly but trickled down the sides of the lizards' heads.

26.—Phrynosoma platyrhinos Girard. DESERT HORNED TOAD.

Phrynosoma platyrhinos, GIRARD, Stansbury's Exped. Gt. Salt Lake, 1853, pp. 361, 363, pl. VII, figs. 1–5 (type locality **Great Salt Lake**); STEJNEGER, N. A. Fauna, No. 7, 1893, p. 190, pl. II, figs. 4a–4o.

Doliosaurus platyrhinos, GIRARD, U. S. Explor. Exped., Herp., 1858, p. 409.

Anota calidiarum, COPE, Am. Nat., XXX, No. 358, Oct., 1896, p. 833 (type locality "**Death Valley, Cal.**" [uncertain]).

Description.—Nostrils opening above lines joining superciliary ridges with end of snout. Head-spines of moderate size or rather short; five to seven temporals, one occipital, and one or two postorbitals, on each side. Three scales in front of occipital horns much larger than other head-shields. Latter usually almost flat, except just in front of occipital and temporal spines, but roughened with small ridges and granulations. Gular region covered with small granular scales, either uniform or with one series of larger scales at each side. Below lower labials, and separated from them by one or two rows of small scales, is a series of large spinose plates which increase in size posteriorly. Two groups of weak spines on each side of neck, lower somewhat larger than upper. Back, tail, and upper surfaces of thighs bearing scattered, slightly elevated, keeled, tubercular scales, with smaller scales and granules between. A single series of peripheral spines, gradually disappearing posteriorly. Tail edged with a row of small spines. Scales on front of the arm large, pointed, and strongly keeled. Those on chest, abdomen, and proximal half of tail smooth. Tympanum usually covered with scales, but sometimes naked. Femoral pores varying from seven to twelve on each side, often invading preanal region. Males with enlarged postanal plates.

The general color of the upper surfaces is white, gray, yellow, brown, or red, variously marbled with black, brown, or slate. A large dark area on each side of the neck is much more distinct in young than in adults. The usual dark dorsal blotches are very indistinct, as are also the dusky cross-bands on the tail. The head is usually dotted with black or brown. The lower surfaces are yellowish white, uniform, or spotted with black, brown, or slate.

Length to anus	30	38	48	77	85	94
Length of tail	14	29	22	40	45	46
Snout to ear	7	8	10	15	16	16
Width of head*	9	12	14	21	22	23
Length of occipital horn	2	2	3	6	8	8
Fore limb	16	19	22	35	34	37
Hind limb	20	25	30	46	48	52
Base of fifth to end of fourth toe	8	9	10	14	15	16

Distribution.—The range of this horned toad in California includes the Colorado and Mojave Deserts of San Diego, Riverside, San Bernardino, and Kern Counties, and the hot, arid portions of the Great Basin. It may, perhaps, be found also in the northeastern corner of the State, for it has been reported from Oregon and is very common on the plains near the Snake River in Idaho. It has not been taken in California anywhere west of the deserts.

It crosses Nevada (Pyramid Lake, Ash Meadows, Amargosa, Vegas Valley, Pahrump Valley, Pahranagat Valley, Indian Spring Valley, Panaca, Grapevine Mts.) to western Utah.

Habits.—Like other species of this genus, *Phrynosoma*

---

*To tips of the temporal horns.

platyrhinos feeds upon small insects. These it catches upon the ground and rarely if ever attempts to climb. It cannot run swiftly, but sometimes tries to escape by burying itself in the loose desert soil.

27.—Phrynosoma m'callii (Hallowell). FLAT-TAILED HORNED TOAD.

> *Anota M'Callii,* HALLOWELL, Proc. Acad. Nat. Sci. Phila., VI, 1852, p. 182 (type locality "Great Desert of the Colorado, between **Vallicita and Camp Yuma, about 160 miles east of San Diego** "); HALLOW., Sitgreave's Zuni and Colorado Rivers, 1853, p. 127, pl. 10.
>
> *Doliosaurus mc'calli,* GIRARD, U. S. Explor. Exped., Herp., 1858, p. 408; BAIRD, U. S. Mex. Bound. Surv., Rept., p. 9, pl. 28, figs, 4–6.

Description.—Snout very short, with nostrils opening above continuations of superciliary ridges. Large head-spines; one slender occipital, three to five temporals, and five to seven sublabials, on each side. Sometimes a small interoccipital horn. Scales on upper surface of head slightly convex and nearly smooth, two on occiput being largest. Supralabials small, but projecting, making margin of upper lip serrate. Gular region covered with small, smooth scales, of which one series on each side is slightly enlarged. Below infralabials a series of very large, spinose plates. Two or three small groups of spines on sides of neck. Back, tail, and upper surfaces of thighs bearing scattered, very slightly elevated, weakly keeled, tubercular scales, with small keeled scales or smooth granules between. Two or three series of peripheral spines; those of upper or of middle series largest. Tail greatly flattened and bearing a fringe of thickly set, slender spines. Scales on front of arm large, pointed, and strongly keeled. Those on chest small and smooth, except anteriorly, where larger and keeled. Scales of abdomen small and smooth. Most of lower caudal scales keeled. Tympanum entirely covered with granular scales. Femoral pores arranged in long series, eighteen or twenty on each side.

The body is ash-color or yellowish olive above, with a narrow median dorsal line of black or dark brown, extending from the occiput to the base of the tail. There is a brown blotch on each side of the neck. Double series of rounded dark spots ornament each side of the back, uniting to form faint cross-bars on the tail. The lower surfaces are silvery or yellowish white.

Length to anus	48
Length of tail	22
Snout to ear	9
Width of head	17
Length of occipital spine	6
Fore limb	23
Hind limb	29
Base of fifth to end of fourth toe	10

Distribution.—The locality in San Diego County at which the original specimen of this species was secured is stated as the Great Desert of the Colorado, between Vallecita and Camp Yuma, about one hundred and sixty miles east of San Diego. I believe that this is the only Californian record of the Flat-tailed Horned Toad.*

Family IV. ANGUIDÆ.

In the lizards of this family, the tongue is formed of a larger, thick, posterior portion, and a smaller, thin, emarginate, anterior part which is more or less retractile into a fold of the posterior portion. The imbricate scales are reinforced with bony plates. In some genera the limbs are well developed, but in others they are

---

* Yarrow and Henshaw record several specimens from the Mojave Desert as belonging to this species (Surv. W. 100th Mer., Append. NN, 1878, p. 225), but these, doubtless, were really *P. platyrhinos.*

rudimentary or even absent. The family is represented in California by a single genus.

Genus 14. GERRHONOTUS.

Gerrhonotus, Wiegm., Isis, 1828, p. 379; *Abronia*, Gray, Ann. &
Mag. Nat. Hist., I, 1838, p. 389; *Elgaria*, Gray, l. c., p. 390.

There are four pentadactyle limbs. The head and body are elongate, but shorter than the tail. The head-plates are rather large and change gradually to those of the neck. Interoccipital and azygous prefontal plates are present. The dorsal, caudal, and ventral scales are large, rhomboidal, and arranged in transverse as well as longitudinal series. A band of granules along each side of the body is usually hidden by a dermal fold. The eye is large, with round pupil and well developed lids. The ear-opening is distinct. There is no transverse gular fold. Femoral and preanal pores are absent.

SYNOPSIS OF SPECIES.

a.—Lower temporal scales smooth.
  b.—Dorsal and caudal scales strongly keeled.
    c.—Dorsal scales in fourteen (rarely 14⅔ or 12⅔) longitudinal series; *
      interoccipital plate (usually) single; azygous prefrontal large and
      very broad; belly rarely without gray or slate-colored longitudinal
      lines, which, when present, run along the middle of each series of
      ventral plates.†.................... G. scincicauda.—p. 103.
    c².—Dorsal scales in sixteen (rarely 15¼ or 18) longitudinal series; in-
      teroccipital usually represented by two to four small scales;
      azygous prefrontal usually of moderate size or small; gray or
      slate-colored longitudinal lines between the series of ventral
      plates, if present................... G. burnettii.—p. 107.

*The scales of the row nearest the granular area vary somewhat in size in different
specimens. When counting the dorsal series, the lowest (on each side) should not be in-
cluded if its scales are less than half the size of those immediately above them. When
its scales are half the size of those above, I have called the lowest row ½ series; when
more than half the size of those above, a whole series.

† Specimens of *G. scincicauda* and *G. burnettii* can usually be recognized at a glance, but
the amount of individual variation is so great that it is very difficult to express their
characteristics in a key which will serve to distinguish all specimens. It should be re-
membered that single specimens may vary in one or more of the characters given (except

b². —Dorsal and caudal scales weekly keeled.

Dorsal scales in fourteen (or 14½) longitudinal series; interoccipitals two or three (rarely one); azygous prefrontal of moderate size or small; dark ventral lines between the series of scales, if present; back without complete dark cross-bands. **G. principis.**—p. 112.

a². —Temporal scales keeled.

Dorsal scales in sixteen longitudinal series; interoccipital plate single; azygous prefrontal large; back without distinct dark cross-bands .... **G. palmeri.**—p. 113.

28.—Gerrhonotus scincicauda (Skilton). ALLIGATOR LIZARD.

? Gerrhonotus Weigmannii, GRAY, Cat. Liz. Brit. Mus., 1845, p. 54 (type locality **Mexico?**); O'SHAUGHNESSY, Ann. Mag. Nat. Hist. (4), XII, 1873, p. 46.

Tropidolepis scincicauda, SKILTON, Am. Journ. Sci. Arts, (2), VII, 1849, p. 202, pl. at p. 312, figs. 1–3 (type locality **Dalles of the Columbia**).

Elgaria scincicauda, BAIRD & GIRARD, Stansbury's Exped. Gt. Salt Lake, 1853, p. 348, pl. IV, figs. 1–3; GIRARD, U. S Explor. Exped., Herp., p. 210, pl. XXIII, figs. 1–9.

the position of the dark ventral lines), which are based upon an examination of 150 specimens. Some of the details of these specimens are:

	scincicauda.	*burnettii*
Number with 12 2-2 scale rows	3	..
" " 14 " "	80	..
" " 14 2-2 " "	3	..
" " 15½ " "	..	4
" " 16 " "	..	59
" " 18 " "	..	1
	86	64
Number with one interoccipital	77	7
" " 2 to 4 "	8	59
" " no "	1	1
	86	63
Number with azygous prefrontal large	82	7
" " " " moderate	4	20
" " " " small	..	34
	86	61
Number without dark lines on belly	2	16
" with dark lines in middle of ventral scale rows	84	..
" " " " at junction " " " "	..	48
	86	64

Elgaria grandis, BAIRD & GIRARD, Proc. Ac. Nat. Sci. Phila., VI,
1852, p. 176 (type locality **Oregon**); GIRARD, U. S. Explor.
Exped., Herp., p. 212, pl. XXII, figs. 1-8.

Gerrhonotus multicarinatus, YARROW, Bull. U. S. Nat. Mus., No. 24,
1882, p. 47 (part); COPE, Proc. Ac. Nat. Sci. Phila., 1883, pp.
29, 32.

Gerrhonotus cæruleus, BOULENGER, Cat. Liz. Brit. Mus., II, 1885,
p. 273 (part).

Gerrhonotus scincicauda, STEJNEGER, N. A. Fauna, No. 7, 1893,
p. 195.

Description.—Body long and rather slender, with
short limbs and very long tail. Head pointed, with
flattened top and nearly vertical sides; its temporal
regions often greatly swollen in old specimens. Rostral
plate rounded in upper outline. Behind it, on top of
the head, a pair of small internasals, a pair of small
frontonasals (sometimes absent), a very large azygous
prefrontal, a pair of large prefrontals, a long frontal, a
pair of frontoparietals, two parietals separated by an
interparietal, a pair of occipitals, and a (usually) single
interoccipital. Two series (of 5 & 3) supraoculars and
a series of small superciliaries. Upper temporal scales
usually keeled, but lower two or three series smooth.
Upper labials much larger than lower. Below latter two
series of large sublabial plates, lower larger. Gular
scales smooth and imbricate. Scales on upper surfaces
and sides of neck, body, and tail large, rhomboidal,
slightly oblique, strongly keeled, strengthened with
bony plates, and arranged in both transverse and longitudinal series. Number of longitudinal dorsal series on
body fourteen (rarely 12⅔ or 14⅔). Number of transverse series between interoccipital plate and backs of
thighs varying from forty-one to fifty-two (average in
85 specimens, 47.6). A band of granules along each
side from the large ear-opening to the anus, usually

hidden by a strong fold.* Ventral plates about size of dorsals, smooth, imbricate, and arranged in twelve longitudinal series. Number of scales between sym-physeal plate and anus varying from sixty-two to sixty-eight.

The ground color above, in adults, is olive, brown, yellow, red, or gray, usually paler on the sides and crossed, on the neck and body, by from nine to sixteen continuous irregular black or dark brown bands. These bands are usually of about the width of one row of scales, but are undulate and sometimes more or less diffused on the back. The lateral scales which these bands occupy are tipped with white. Sometimes the tail is marked like the back, but often it bears merely a central row of small brown blotches. The head and

---

* This fold often disappears in specimens full of eggs or food.

limbs may be either unicolor or irregularly mottled with brown. The lower surfaces are white or yellowish, sometimes suffused with pale brown or gray. The abdominal and thoracic regions are rarely without gray or slate-colored lines along the middle of each longitudinal series of scales.

Young specimens are at first indistinguishable from *G. burnettii* of a similar age, but the complete dorsal cross-bands very soon appear.

Length to anus	41	60	80	114	135	154
Length of tail	83	139	159	226	294	
Snout to ear	10	12	17	23	30	34
Width of head	7	8	11	15	21	28
Head to interoccipital	9	11	14	18	22	25
Fore limb	12	14	20	30	36	40
Hind limb	16	21	27	39	48	53
Base of fifth to end of fourth toe	5	7	10	14	16	17

Distribution.—The Alligator Lizard ranges over the whole length of California, but, I believe, has never been found east of the Sierra Nevada or on the Colorado or Mojave Deserts. In the south it appears to be the only species of the genus, but in the north its range overlaps that of *G. burnettii, G. palmeri,* and *G. principis.* Mr. H. W. Henshaw found it on Santa Cruz Island. I have examined specimens from Santa Rosa Island and from Shasta (Redding), Mendocino (Irishes, Fairbanks), Lake (Blue Lakes, Kelseyville), Napa (St. Helena), Sonoma (Santa Rosa), Marin (Lagunitas), Alameda (Calaveras Valley, Livermore, Haywards, Oakland, Berkeley), Santa Clara (Palo Alto, Santa Clara, College Park, Smith Creek, Los Gatos), Santa Cruz (Corralitos, Soquel), Monterey (Pacific Grove), El Dorado (Fyffe 3700 feet, Riverton 4000 feet), Mariposa (near Wawona), Tulare (Three Rivers, East Fork Kaweah), Santa Barbara (Santa Barbara), San Bernardino (Lytle Creek),

Riverside (San Jacinto, Riverside), and San Diego (Carlsbad, Cuyamaca Mountains) Counties, California, and Douglass (Drain) County, Oregon. It seems to be most abundant in the chaparral country, but is by no means confined to this belt.

Habits.—This large and elegantly marked species is rather slow of movement, but its sluggishness is largely due to its lack of timidity, for, if thoroughly frightened, it sometimes runs with great swiftness. It is usually to be seen on the ground, but frequently climbs through the bushes. At such times its long prehensile tail must be very useful. Its food is made up chiefly of insects, such as beetles and flies. Like the following species *(G. burnettii)*, the Alligator Lizard is' ovoviviparous. Messrs. Doane and Ely brought me a pair which they found mating in a bush near Palo Alto, May 12, 1894. This lizard sometimes bites fiercely when caught, but, like all Californian reptiles excepting the rattlesnakes, is not poisonous.

29.—Gerrhonotus burnettii Gray. BURNETT'S ALLIGATOR LIZARD.

> ? *Gerrhonotus cæruleus*, WIEGM., Isis, 1828, p. 380 (type locality " Brazil '"*); WIEGM., Herp. Mex., 1834, pp. 29, 31; BOCOURT, Miss. Sci. au Mex., Rept., 1878, p. 353, pl. XXIc, figs. 3, 3a; BOULENGER, Cat. Liz. Brit. Mus., II, 1885, p. 273 (part).
>
> *Gerrhonotus burnettii*, GRAY, "Griff. An. King., IX., Synop. Rept., 1831, p. 64" (type locality "South America"); GRAY, Ann. Mag. Nat. Hist., I, 1838, p. 390; GRAY, Beechey's Voy., Zool., 1839, p. 96, pl. XXXI, fig. 2; GRAY, Cat. Liz. Brit. Mus., 1845, p. 54; O'SHAUGHNESSY, Ann. Mag. Nat. Hist. (4), XII, 1873, p. 47; BOCOURT, Miss. Sci. au Mex., Rept., 1878, p. 356, XXIc, figs. 4, 4a; STEJNEGER, N. A. Fauna, No. 7, 1893, p. 197.

---

*Bocourt quotes from Peters (Miss. Sci. au Mex., 5e liv., p.355): "Ce Gerrhonote a été rapporté par M. Chamisso, que a fait des collections sur les côtes occidentales des deux Amériques, aussi se pourrait-il qu'il ait été recueilli en Californie, d'où ce voyageur a rapporté divers objets d'histoire naturelle."

Elgaria formosa, BAIRD and GIRARD, Proc. Ac. Nat. Sci. Phila., VI,
1852, p. 175 (type locality **California**); GIRARD, U. S. Explor.
Exped., Herp., p. 206, pl. XXIII, figs. 10–17.

Description.—Body long and rather slender, with short
limbs and very long tail. Head pointed, with flattened
top and nearly vertical sides, its temporal regions sometimes swollen. Rostral plate large, and rounded in
upper outline. Behind it, on top of head, a pair of
small internasals, a pair of frontonasals, a small or
moderate-sized azygous prefrontal, a pair of prefrontals,
a large frontal, a pair of frontoparietals, two parietals
with an interparietal between them, and a pair of occipitals separated by from one to four, usually by two or
three, interoccipitals. Two series of (5 and 3) supraoculars and a series of small superciliaries. Upper
temporal scales usually keeled, but lower two or three
series smooth. Upper labials much larger than lower.
Below latter, two series of large sublabial plates, lower
much the larger. Gular scales imbricate and smooth.
Scales on upper surfaces and sides of neck, body,
and tail rhomboidal, slightly oblique, strongly keeled,
strengthened with bony plates, and arranged in both
transverse and longitudinal series. Number of longitudinal series on body sixteen (rarely $14\frac{2}{3}$ or 18). Number of transverse series between occipital plates and
back of thighs varying from forty-three to fifty-two
(average in 63 specimens, 48.5). A band of granules
along each side from large ear-opening to anus, usually
hidden by a strong dermal fold. Ventral plates about
size of dorsals; smooth, imbricate, and arranged in
twelve (or 13) longitudinal series. Number of scales
from symphyseal plate to anus varying from fifty-eight
to sixty-four.

The ground color above, in adults, is gray, olive,

yellow, green, brown, or almost black, with numerous irregular black or dark brown cross-bands, which, however, usually are broken up into two lateral series of vertical bars and one median series of irregular spots or blotches. The ground color of the longitudinal band

between the median and lateral dark markings is often lighter than elsewhere. Most of the lateral scales occupied by the dark bars are tipped with white. The coloration of the tail is similar to that of the back. The head and limbs may be either unicolor or irregularly mottled with black or brown. The lower surfaces are white, yellow, green, or gray, often with dark gray or slate-colored lines, which, when present, appear between the longitudinal series of scales.

The young are similarly colored, but the dorsal bands are always broken and the medial spots are much smaller than is usual in adults. The ground color of newly born young is an iridescent bronze.

Length to anus...................	27	52	76	88	98	99
Length of tail...................	32	89	126	145	162	172
Snout to ear....................	7	11	15	17	19	20
Width of head	5	7	11	12	14	14
Head to interoccipital..........	6	10	12	14	16	16
Fore limb......................	7	12	18	22	23	24
Hind limb......................	8	16	24	30	31	33
Base of fifth to end of fourth toe.....	3	6	8	11	10	12

Distribution.—Burnett's Alligator Lizard occupies, so

far as is at present known, merely a narrow strip of country extending along the coast from Monterey to Mendocino County, California.* Parts, at least, of this area it holds in common with its larger congener *G. scincicauda.* I have examined specimens from Mendocino (Irishes), Sonoma (Healdsburg), Marin (Mill Valley), San Francisco (Lake Merced, Presidio), San Mateo (Searsville, Pescadero), Santa Clara (Palo Alto), Santa Cruz (Boulder Creek, Big Trees, Glenwood, Soquel), and Monterey (Pacific Grove) Counties.

Habits.—These slow-moving lizards may easily be caught on the sand hills of San Francisco, where they are very common. They are insect-eaters, feeding chiefly upon beetles. Females usually show little resentment when handled, but males often become very angry and will hiss and bite fiercely, although unable to draw blood. A captive male would hiss and jump at my fingers whenever the door of his cage was opened. The skin is renewed, sometimes at least, twice each year, and, contrary to the method usual among lizards, is shed in a single piece, the animal escaping, as it were, through its own mouth, and neatly inverting its former covering. The tail is strongly prehensile.

The eggs are retained in the body until the young are fully formed. If numerous, the lateral fold gradually disappears as they increase in size. The young are coiled up in a thin, transparent membrane when born. They almost immediately push the snout through this covering by straightening the body and in the course of a few minutes set themselves entirely free. The number of young varies from two to fifteen, but is usually about seven. Two females were caught June 5,

---

*I have seen two typical specimens said to have been collected in the Cuyamaca Mts., San Diego Co., but this locality needs confirmation.

1895, and put in small cages, where there were supplied with flies and water, of which these lizards are very fond. Young appeared in one box August 29 and in the other September 24, 1895. Those of the first brood varied in length from seventy-one to seventy-six millimeters, and those of the second, from fifty-eight to sixty-two. The old lizards showed no affection or solicitude for their young, but the young liked to be near their parents. Six out of fifteen inherited an irregularity of the dorsal scale-series, shown by their female parent.*

During the first few days these young lizards ate nothing, but then they began to snap at the smaller flies. When stalking flies, they crouched close to the ground and crept slowly forward, their heads swaying from side to side and their tails quivering or thrashing with excitement. Then, if the snap was successful, the prey was held firmly in the jaws while the lizard, with body and tail straightened, rolled rapidly over and over, grinding the fly in the sand. Frequently when one had caught a fly the others would rush up and feel of it inquisitively with their tongues, sometimes even trying to appropriate it to themselves. Sometimes, too, one's chase was interrupted by another lizard seizing the quivering tip of the hunter's tail. The young lizards were very fond of lying in the water, and several deliberately held their heads under its surface until they were drowned. The last of the family died, May 5, 1896, during a vain endeavor to shed its skin.

The lizards which I kept in confinement were more or less active throughout the winter, but Mr. James M. Hyde broke up two decaying logs, near Pescadero, December 22, 1893, and found five lizards of this species

---

* I have found a similar irregularity in only two of forty-nine other specimens. One of these was from the same locality as this female.

hibernating with five *Sceloporus occidentalis* and one
Eumeces skiltonianus.

30.—Gerrhonotus principis (Baird & Girard). NORTHERN ALLIGATOR LIZARD.

Elgaria principis, B. & G., Proc. Ac. Nat. Sci. Phila., 1852, p. 175
(type locality **Oregon and Puget Sound**); GIRARD, U. S.
Explor. Exped., Herp., p. 214, pl. XXII, figs. 9–16.

Description.—Body long and rather slender, with
short limbs and long tail. Head pointed, with flattened
top and almost vertical sides, its temporal regions sometimes slightly swollen. Rostral plate large, and rounded
in upper outline. Behind it, on top of head, follow a
pair of small internasals, a pair of frontonasals, a moderate-sized or small azygous prefrontal, a pair of prefrontals, a long frontal, a pair of frontoparietals, two
parietals with an interparietal between them, and a pair
of occipitals separated by one or usually two or three
interoccipitals. Two series of supraoculars and a series
of small superciliaries. Upper temporal scales usually
smooth and lower two or three series always so. Upper
labials much larger than lower. Below latter, two series
of sublabial plates, lower much the larger. Gular scales
imbricate and smooth. Scales on upper surfaces and
sides of neck, body, and tail large, rhomboidal, slightly
oblique, weakly keeled, strengthened with bony plates,
and arranged in both transverse and longitudinal series.
Number of longitudinal series on body fourteen (rarely
$14\frac{2}{2}$ or 16). Number of transverse series between occipital plates and back of thighs varying from forty-four
to fifty-three. A band of granules along each side from
large ear-opening to anus, usually hidden by a strong
dermal fold. Ventral plates about size of dorsals,
smooth, imbricate, and arranged in twelve longitudinal

series.   Number of scales from symphyseal plate to anus varying at least from fifty-six to sixty-two. The ground color above is olivaceous brown, without cross-bands, but with numerous irregular dark brown spots, which sometimes form longitudinal series. The head and limbs are usually more or less clouded with dark brown. The lower surfaces are yellowish or greenish white, often slightly washed with gray, and with or without slate-colored lines between the longitudinal series of scales.

Length to anus...................... 90	91	96	96	102	105
Length of tail......................141		151	139	152	148
Snout to ear..................... 16	17	18	18	19	19
Width of head 12	12	12	12	13	13
Head to interoccipital............... 14	14	15	15	15	15
Fore limb........................ 21	21	23	22	23	23
Hind limb........................ 28	27	29	29	29	29
Base of fifth to end of fourth toe...... 10	10	11	10	11	11

Distribution.—The Northern Alligator Lizard is a species of western Washington and Oregon, whose range seems to extend south on the western slope of the Sierra Nevada of California at least as far as Red Point, Placer County.* It may, perhaps, live also in the northwestern corner of the State. I believe that this form will be found to intergrade with both *G. burnettii*, of the coast, and *G. palmeri*, of the southern Sierra Nevada. Northward its range extends to Vancouver Island. It is very abundant near Puget Sound.

31. — Gerrhonotous palmeri (Stejneger). MOUNTAIN ALLIGATOR LIZARD.

Gerrhonotus scincicauda palmeri, STEJNEGER, N. A. Fauna, No. 1893, p. 196 (type locality South Fork King's River, Calif.).

Description.—Body long and rather slender, with

---

* The identity of this specimen is not certain. It may, possibly, be an abnormal example of *G. palmeri.*

short limbs and long tail. Head pointed, with flattened
top and nearly vertical sides; its temporal regions somewhat swollen. Large rostral plate rounded in upper
outline. Behind it, on top of head, follow a pair of
small internasals, a pair of small frontonasals, a large
azygous prefrontal, a pair of large prefrontals, a long
frontal, a pair of frontoparietals, two parietals separated
by an interparietal, and a pair of occipitals with one or
more interoccipitals between them. Two series (of 5
and 3) supraoculars and one series of small superciliaries. All temporal scales keeled. Upper labials much
larger than lower. Below latter, two series of sublabial
plates, interior larger. Gular scales imbricate and
smooth. Scales on upper surfaces and sides of neck,
body, and tail large, rhomboidal, slightly oblique,
strongly keeled, reinforced with bony plates, and arranged in both longitudinal and transverse series.
Number of longitudinal series on body sixteen. Number of transverse rows between interoccipital plate and
backs of thighs varying from forty-two to forty-nine in
specimens examined. A band of granules along each
side from large ear-opening to anus, usually hidden by
a strong dermal fold. Ventral plates about size of dorsals, smooth, imbricate, and arranged in twelve (or 13)
longitudinal series. Number of scales from symphyseal
plate to anus fifty-nine to sixty-two.

The ground color above is olive-brown or bluish or
greenish drab, usually a little paler laterally than near
the middle of the back. There are no definite cross-
bands, the dark pigments appearing in ill-defined marblings or blotches on the back, or in white-tipped black
spots on the sides. The head and limbs are usually
unicolor, but may be marked with darker brown. The
lower surfaces are yellowish or greenish white, some-

times slightly washed with gray. There are no definite longitudinal lines on the belly in the specimens which I have seen, but two specimens have indications of them between the rows of scales.

Length to anus 37	88	91	98	105	120
Snout to ear.............,.........:......... 9	19	19	22	23	26
Width of head........................ 6	14	14	17	18	21
Head to interoccipital................. 8	15	16	18	19	20
Fore limb............................10	24	24	25	31	32
Hind limb..... 14	33	33	37	36	43
Base of fifth to end of fourth toe 5	12	12	14	14	15

Distribution.—The Mountain Alligator Lizard has been found only at high altitudes (5,000 to 9,000 feet) on the western slope of the Sierra Nevada of Tuolumne, Mariposa, Fresno, and Tulare Counties, and at a slightly lower level (Fyffe 3,700 feet) in El Dorado County. It has been recorded from various localities along the Kern, Kaweah, and King's Rivers and is common in the Yosemite Valley.

Habits.—This species is common near the Little Kern River. Here it hides behind the loose bark of the great pines. Like other members of the genus, it usually moves slowly and seems to have much curiosity. Near the Yosemite Valley it mates about the middle of June.

Family V.   ANNIELLIDÆ.

This family, which is confined to California, contains a single genus of strongly degraded lizards. The body is cylindrical and snake-like, without strongly marked neck or tail. There are no external traces of limbs, but a rudimentary pelvis remains. The tongue is thick, with a thinner, smooth, deeply notched anterior portion. The teeth are few, but large and curved. Thin osteodermal plates are present.

Genus 15.   ANNIELLA.

Anniella, GRAY, Ann. & Mag. Nat. Hist. (2), X, p. 440 (type pulchra).

The scales are small, smooth, imbricate, and rather soft; the dorsals, laterals, ventrals, and caudals nearly equal-sized. The ears are entirely concealed and the eyes partially so. The tail is very blunt and ends in a round plate. The preanal scales are numerous. The head-plates are few and large. The nasal extends to or almost to the labial margin, the first labial appearing on the lower surface of the lip.

SYNOPSIS OF SPECIES.

a.—Color above drab or silvery gray or yellowish white, with three or more black or brown lines.....................**A. pulchra.**—p. 116.

a².—Color above black or blackish brown, with or without dark longitudinal lines................................**A. nigra.**—p. 118.

32.—Anniella pulchra Gray. SILVERY FOOTLESS LIZARD.

Anniella pulchra, GRAY, Ann. Mag. Nat. Hist. (2), X, 1852, p. 440 (type locality **California**); "GRAY, Zool. Herald, p. 154, pl. XXVIII"; BOCOURT, Miss. Sci. au Mex., p. 460, pl. XXIIG, fig. 2; BAUR, Proc. U. S. Nat. Mus., XVII, 1894, p. 345.

Description.—Head slightly depressed, rather short, scarcely distinct from neck even in old examples where temporal regions have become swollen. Snout projecting beyond lower jaw. Rostral plate very large and strongly recurved on top of snout, where separated from frontal by a pair of large prefrontals. Behind large frontal, a single very broad frontoparietal, its posterior margin notched to receive a small interparietal with which it frequently unites. On each side of interparietal, a small parietal, and behind these usually two small occipitals separated by an interoccipital. A large supraocular precedes several smaller plates. A large preocular with, usually, a smaller one below it. Nasal large and extending to margin of lip, but small first supralabial plate may be seen below it. Second supralabial largest. Symphyseal large, followed by several pairs of large sublabials. Infralabials smaller than supralabials. Dorsal, ventral, lateral, and caudal scales all similar, slightly largest on tail and smallest on neck, strongly imbricate, rounded in posterior outline, and perfectly smooth. Preanal scales slightly enlarged. Number of longitudinal series of scales around body varying from twenty-four to thirty-four.

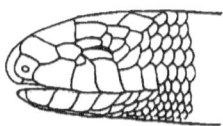

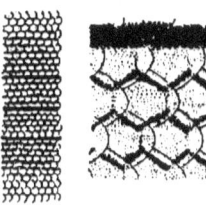

The color above is yellowish white or silvery or drab-gray, with one distinct longitudinal brown line down the middle of the back and one or more similar lines along each side. Very narrow brown zigzag lines usually run along the margins of the other series of dorsal scales. These lines are sometimes quite yellowish, sometimes nearly black. The lower surfaces are yellowish white, frequently suffused with brown, slate, or gray on the chin, throat, and tip of tail, and often showing narrow zigzag longitudinal lines. The entire upper surface of a specimen from San Bernardino is slightly suffused with olive-gray.

Length to anus...................... 84	97	125	130	143	146
Length of tail...................... 44	59	70	74	89	96
Width of head...................... 4	4	5	5	6	6
Head to interparietal............... 4	5	5	5	5	6
Diameter of body................... 4	5	5	7	7	7

Distribution.—The most northern localities from which I have obtained specimens of this lizard are San Ardo, in the interior of Monterey County, and Bear Valley, in San Benito County. It doubtless occurs in many parts of the San Joaquin Valley, where it has been taken in Tulare County. Farther south it has been found at San Bernardino, and is very common near San Jacinto, Riverside County, and in the western portion of San Diego County.

Habits.—The habits and food of the Footless Lizard or "Silver Snake" are the same as those of *Anniella nigra.*

33.—**Anniella nigra** Fischer.   BLACK FOOTLESS LIZARD.

Anniella nigra, FISCHER, Abh. Nat. Vereiu Hamburg, IX, 1, 1885 (1886), p. 9 (type locality **San Diego, California**).

Description.—Head very slightly depressed, short, and scarcely distinct from neck. Snout projecting beyond lower jaw. Rostral plate very large and strongly recurved on top of snout, separated there from frontal by a pair of large prefrontals. Behind large frontal, single very broad frontoparietal, its posterior margin notched to receive a small interparietal with which it sometimes unites. On each side of interparietal, a small parietal, and behind these usually two small occipitals separated by an interoccipital. One large and one or more small supraoculars and a series of small superciliaries. Large preocular with a smaller one below it. Nasal large and extending to margin of lip, but a small first supralabial may be seen below it. Second supralabial largest. Symphyseal large, followed by several pair of large sublabials. Infralabials smaller than supralabials. Dorsal, lateral, ventral, and caudal scales all similar, slightly largest on tail and smallest on neck, strongly imbricate, rounded in posterior outline, and perfectly smooth. Preanal scales slightly enlarged.

Number of longitudinal series of scales around body not differing from *Anniella pulchra*. The entire upper surface is deep blackish brown, with or without indistinct lines of darker brown or black corresponding in position with those of *A. pulchra*. The chin, throat, and the tip of the tail are suffused with dark brown. The rest of the lower surface is yellowish white, sometimes with narrow brown zigzag lines between the longitudinal series of scales.

Length to anus	117	137	148	149	149	161
Length of tail	17*	68	26*	17*	17*	20*
Width of head	5	5½	6	6	7	7
Head to interparietal	4	5	5	6	6	6
Diameter of body	5	6	7	7	8	8

If the type of *A. nigra* really came from San Diego it is doubtful if this form is worthy of recognition. However, it seems best to retain it here, because all the numerous black specimens which I have seen were collected on the coast of Monterey County, where no specimen of the light form has been found.

Distribution.—Pacific Grove, Monterey County (and "San Diego, California"?).

Habits.—The Black Footless Lizard burrows in the soil of the pine forest at Pacific Grove. It is sometimes found under stones or boards, but travels swiftly under the surface of the loose soil. An examination of the contents of several stomachs has shown its food to consist of large insect larvæ (more than 1¼ inches long), and two small ground dwelling beetles *(Helops* and *Platydema)*. Mr. Harold Heath, of Stanford University, has found this lizard to be ovoviviparous.

---

* Reproduced??

Family VI.   HELODERMATIDÆ.

This family contains the only lizards which are known to be poisonous. There is but a single genus, with two species. The tongue is large, deeply divided at tip, smooth anteriorly but villose posteriorly. The teeth differ from those of other lizards in being grooved. There are large poison-glands under the chin. The limbs are well developed. The skin of all the upper surfaces is covered with large tubercles which often ossify. The belly is provided with squarish plates. Usually there are no femoral or preanal pores, but one specimen has a single preanal pore of great size.

Genus 16.   HELODERMA.

Heloderma, Wiegm., Isis, 1829, p. 624 (type horridum).

There are four pentadactyle limbs. The head is covered with irregular, convex, bony plates, which often coössify with the skull. The back and sides are provided with more or less regular rows of tubercles similar to those on the head. The ventral plates are arranged in transverse series. The eye has well developed lids and a round pupil. The ear-openings are large. One strong and usually one or more weaker gular folds are present.

34.—Heloderma suspectum Cope.   GILA MONSTER.

Heloderma suspectum, COPE, Proc. Ac. Nat. Sci. Phila., 1869, p. 5 (type locality Sonoran Region); SHUFELDT, Proc. Zool. Soc. Lond., 1890, p. 148; STEJNEGER, N. A. Fauna, No. 7, 1893, p. 194.

Description.—Head and body depressed, large, heavily built, with short limbs and tail. Upper surface of head broad, flat, and covered with large, irregular, convex, bony tubercles. Snout rounded. Temporal regions swollen. Nostrils large, opening laterally between three plates. Eye rather small. Ear-opening large, elliptical, oblique, and overhung by temple. Rostral and symphyseal plates large. A pair of internasals. Three pair of plates behind symphyseal. Gular region and fold with small round or oval, convex, or flattened tubercles, changing gradually into the plates of the belly. Body, limbs, and tail covered above and laterally with nearly equal-sized, round, smooth, convex tubercles separated by granules. Lateral tubercles passing gradually into smooth, flat, squarish plates in transverse rows on lower surfaces of body and tail. Digits with transverse plates above and below. A pair of enlarged plates in front of anus.

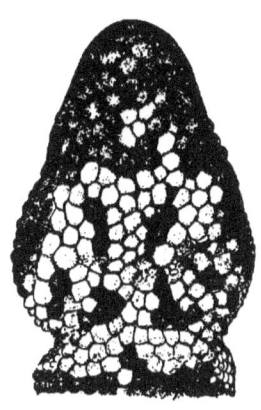

Probably no two specimens show just the same pattern of coloration. The top of the head, the body and limbs are variously marbled, banded, or reticulated with orange or salmon and black or brown. The chin, throat, snout, and sides of head are usually of the dark color with few if any orange or salmon-colored tubercles. The markings on the tail frequently form transverse bars or rings. The belly is orange or salmon and black or brown, tessellated.

Length to anus....................220	270	288	295	315	345
Length of tail.....................101	125	140	145	140	150
Snout to ear......................... 36	46	53	57	53	58
Width of head..................... 31	43	50	49	49	52
Fore limb......................... 62	78	80	83	87	93
Hind limb......................... 66	74	90	88	95	98
Base of fifth to end of fourth toe..... 18	20	23	25	25	25

Distribution.—The Gila Monster has been found in the Valley of the Virgin, about eight miles below Bunkerville, near the eastern boundary of Nevada. It may be that it occurs on portions of the deserts of southeastern California, but as yet no specimens from this area have found their way into museums.

Habits.—The Gila Monsters are the only lizards whose bite is known to be poisonous. The venom is secreted by large glands situated just under the chin, and flows out, onto the floor of the mouth, between the lips and the gums. Being below the teeth and not directly communicated to them, the poison sometimes fails to find its way into a wound although the teeth are grooved to afford it a passage. The upper jaw of the Monster is provided with a saliva which possesses no poisonous properties. This harmless saliva appears to be present in the lower jaw as well as the upper, but is there mixed with venom about as deadly as that of the rattlesnakes. Although provided with so powerful a poison, the Gila Monster is so gentle and sluggish that it is not always easy to cause one to bite, but when thoroughly angered it bites fiercely, throwing its head to one side with lightning-like quickness and holding like a bull-dog to whatever it has seized. Sumichrast says that it turns onto its back before biting. Although this observation has not been confirmed, the presence of venom in. the lower jaw only would explain such an action. In spite of its clumsy form it sometimes climbs bushes, probably in search of bird's eggs, which, together with young rabbits, etc., form its food.

Family VII.   XANTUSIIDÆ.

This family contains but three genera; one Central American, one West Indian, and one Californian. The

eyes are without lids.   The head is covered with large
shields.   The upper surface of the body is granular or
tubercular, but the lower is provided with plates.   The
tongue is broad, plicate, with tip indistinctly notched.
The ear-opening is large.   Femoral pores are present.

Genus 17.   XANTUSIA.

Xantusia, BAIRD, Proc. Ac. Nat. Sci. Phila., 1858, p. 255 (type
vigilis); *Zablepsis*, COPE, Am. Nat., XXIX, 1895, p. 758 (type
henshawi); *Amœbopsis*, COPE, l. c., p. 758 (type gilberti).

The dorsal granules are uniform.   Superciliary and
sometimes supraocular plates are present.   The interparietal is separated from the frontal by the frontoparietal plates.   The pupil is vertically elliptic.   There
are two or three transverse gular folds, the last edged
with enlarged plates.

SYNOPSIS OF SPECIES.

a.—One series of small plates (superciliaries) over eye.
  b.—Ventral plates in twelve longitudinal series; yellow or brown with
    dark spots on single granules...........**X. vigilis.**—p. 123.
  b².—Ventral plates in fourteen longitudinal series; irregularly marbled
    with yellow lines enclosing large dark brown spots.
                              X. henshawi.—p. 128.
a².—Two series of small plates (superciliaries and supraoculars) over eye;
  ventral plates in sixteen longitudinal series. .**X. riversiana.**—p. 130.

35—Xantusia vigilis Baird.   DESERT NIGHT LIZARD.

Xantusia vigilis, BAIRD, Proc. Ac. Nat. Sci. Phila., 1858, p. 255 (type
locality **Fort Tejon, California**); STEJNEGER, N. A. Fauna,
No. 7, 1893, p. 198, pl. III, figs. 1a–1c.

Description.—Body nearly cylindrical, with very short
limbs.   Upper surface of head flattened, curving to-
wards snout.   Three folds on throat, anterior connecting ears and encircling head.   Nostril opening at junction of rostral, internasal, postnasal, and first labial
plates.   Rostral in contact with first labial and internasal plates.   Two internasals followed by a large

subhexagonal frontonasal. Behind this, two prefrontals (in contact), bordered posteriorly by single broad frontal and first superciliary plates. Each of two frontoparietal plates forming sutures with frontal, second, third, and fourth superciliaries, first supratemporal, parietal, interparietal, and its fellow of opposite side. Parietals and very large interparietal bordered behind by two large occipitals. A row of small supratemporal scutes along outer edge of occipital and parietal plates. Two large loreals in contact below with superior labials and above with frontonasal and prefrontal plates. A large postnasal in front of first loreal. A series of small plates, upper of which are superciliaries, usually surrounding eye. Between this ring and larger loreal, two or three small plates. Four or five superior and three or four inferior labials to a point below middle of eye. Eye large, without lids, with vertical pupil. Its diameter contained about twice in distance from end of snout to orbit. Oblique ear-opening with a very weak anterior denticulation. Inferior labials in contact with large sublabials. First pair of latter in contact on median line. Back, sides, upper and posterior surfaces of limbs, and gular regions, covered with subhexagonal granules. A series of large plates along edge of last gular fold. Ventrals quadrate, in twelve longitudinal and twenty-seven to thirty transverse series. Large preanal plates arranged in two rows of two each, sometimes surrounded by a few smaller scales or granules. Tail conical and covered with whorls of smooth, narrow, and transversely convex scales; its length very variable. Six to ten femoral pores forming a series along each thigh.

The ground color in different specimens varies from smoke gray, through many shades of yellow and brown, to clove brown. Scattered granules are dark brown or black. At times these dark granules are so numerous as to become confluent, with a tendency to form longitudinal lines. In other individuals they are scarcely visible. Some specimens have heavy dotting on a very pale ground; in others the dotting is heavy on a dark ground; many show faint dots on a light ground; and several have few dots on a dark ground. A yellowish line usually runs back on the neck from the outer edge of each occipital plate. Two similar lines may sometimes be seen above these. The lower parts are creamy white, sometimes clouded with brown toward the sides. The young average much darker than the adults.

Length to anus	22	37	42	44	47
Length of tail	24	41	61	47	40*
Shielded part of head	6	9	9	9	10
Snout to ear	5½	8	8	8½	9
Snout to anterior gular fold	5½	8	8	8½	9
Snout to posterior gular fold	9	13	14	15	15
Fore limb	7	10½	11	11	12
Hind limb	9½	15	15½	16	17
Base of fifth to end of fourth toe	4	5½	5½	6	6½

Distribution.—The Desert Night Lizard is the most abundant of its class in the territory it has chosen for its home. It seems to be peculiarly dependent upon the presence of tree yuccas. These weird plants grow in each of the localities from which the lizard has been recorded, viz.: Fort Tejon in the Cañada de las Uvas, Kern County, and Hesperia, San Bernardino County, California, and Pahrump Valley, Nevada.

Dr. Charles H. Gilbert and the writer collected specimens near Mojave, Kern County, and found a portion of a cast skin at Victor, San Bernardino County, in November, 1893. In September of the following year,

_____

* Regrown.

the writer found this species common at Mojave and
Hesperia, and secured a single specimen near Cabazon
on the eastern slope of San Gorgonio Pass, Riverside
County, California. The first three of these localities
are situated in the great *Yucca-arborescens* belt, which
extends along the southwestern edge of the Mojave
Desert. The last is in a small and apparently isolated
grove of smaller tree yuccas, seemingly of another
species.

Habits.—About a mile from the station at Mojave
there is a considerable forest of *Yucca arborescens.* The
many trees and wind-broken branches, which lie decaying on the ground, afford a home to numerous colonies
of white ants, scorpions, vicious looking black spiders,
and several species of beetles. In a deep crack of one
of these branches a small lizard was discovered, which,
when caught, proved to be a young *Xantusia vigilis.*
Probably it had not yet learned how to hide from the
day, for I have never seen another undisturbed individual.

The key to their home once discovered, the collection
of a large series of these lizards was merely a matter of
physical exertion. Every fourth or fifth stem that was
examined gave up its *Xantusia,* and in one instance five,
as many as were previously known to collections, were
found under a single tree.

Most of the lizards were found between the bark and
the ground, but many had hidden in the thick clusters
of dead leaves, from which it was very difficult to dislodge
them. When first exposed to the light, they were dark
colored and seemed dazzled for a moment, during
which they made no attempt to escape. They were not
at all sluggish, however, and, if not caught immediately,
made for the nearest cover as fast as their very short

legs would permit. This cover was often the collector, and the little lizards either hid under his shoes, or climbed his legs, sometimes even reaching his shoulders. They showed no desire to enter the numerous holes in the ground about them, or to escape by burrowing. Put into a glass bottle they become very light colored in a few minutes, but began to turn dark again immediately after sundown. Young specimens were numerous and remained dark longer than adults. Many fragments of cast skins were found, but never a whole skin in one place. The stomachs of several individuals contained the wings of some small dipterous insect, the elytra of a little brown beetle, and some small white bodies which resembled spider's eggs.

Several specimens were taken alive to the Leland Stanford Junior University and kept for some months in a large glass jar in which some fine sand and pieces of wood and bark had been placed. At first, they ventured out from their retreat only at dusk unless disturbed, but after a few days they seemed to become more restless, and, urged perhaps by hunger, showed themselves many times each day. At night, when they were always more active, they often climbed to the top of a piece of yucca stem placed upright in the middle of their cage. No desire to burrow was observed. All declined to show any interest in the small beetles and flies, both dead and living, which were placed in the jar, and they finally became greatly emaciated.

Mojave was visited again in the fall of the following year. The specimens were all caught alive and put into a large glass bottle, but were soon killed by the heat, although care was taken to keep them in the shade as much as possible. Count was kept as the lizards were put in the bottle and showed later that several more

were taken out than had been put in. This may have
been due to a mistake in the record, but was more probably caused by the birth of young after capture. The
adults were afterwards carefully examined and three
were found to contain young, showing that the species
is ovoviviparous. One of the three contains two fetuses,
the others have one each. The fetal specimens are
about the size of the young found under the dead
branches. They were taken on the seventeenth and
eighteenth of September.

36.—Xantusia henshawi Stejneger. HENSHAW'S NIGHT
LIZARD.

> *Xantusia henshawi*, STEJN., Proc. U. S. Nat. Mus., XVI, 1893, p. 467
> (type locality **Witch Creek, San Diego County, Califor-
> nia**); VAN DENBURGH, Proc. Cal. Ac. Sci. (2), V, 1895, p 530.
> *Zablepsis henshavii*, COPE, Am. Nat., XXIX, 1895, pp. 758, 860.
> *Xantusia picta*, COPE, Am. Nat., XXIX, 1895, pp. 859, 939 (type
> locality "**Tejon Pass, California**").

Description.—Body greatly depressed, with very short
limbs. Upper surface of head very flat. Three folds
on throat. Nostril opening in a small scute at junction
of rostral, internasal, postnasal, and first labial plates.
Rostral broad and rather low, bounded by first
labial, nasal, and internasal plates. Two internasals followed by a large subquadrate frontonasal, sometimes divided longitudinally; behind this,
two prefrontals, bordered posteriorly by broad frontal
and first superciliary plates. Each of two frontoparietal
plates in contact with frontal, second, third, and fourth
superciliaries, first supratemporal, parietal, interparietal,
and its fellow of opposite side. Parietals and interparietal bordered behind by two large occipitals. One or
more interoccipitals sometimes present. A row of small
supratemporals along outer edge of occipital and parietal

plates. Two large loreals, in contact below with superior labials and above with frontonasal and prefrontal plates. Eye surrounded by a series of small plates, upper five of which are superciliaries. Between this ring and larger loreal two small plates. Five superior and three inferior labials to a point below pupil. Eye large, without lids, and with vertical pupil. Its diameter contained about twice in distance from end of snout to orbit. Ear-opening with a very weak anterior denticulation. Symphyseal plate very long. Inferior labials in contact with large sublabials. First pair of latter in contact on median line. Back, sides, upper, and posterior surfaces of limbs, and gular regions, covered with subhexagonal granular scales. A series of large quadrate plates along edge of last gular fold. Ventrals quadrate, in fourteen longitudinal and thirty-three or thirty-four transverse rows. Preanal plates arranged in three or four rows, the two medial plates of posterior series being largest. Tail conical, somewhat depressed at its base and covered with whorls of smooth scales, which are very narrow and transversely convex. Eight or ten femoral pores forming a series along each thigh.

The ground color above is broccoli brown. On this are numerous large irregular blotches of very dark seal brown, between which run more or less continuous lines of pale yellow. The upper surfaces of the limbs and head are similarly, but less distinctly, marked. The tail is yellow with irregular blotches and half rings of blackish seal brown. The lower surfaces are uniform yellowish white.

Length to anus.. 57	63	65
Length of tail.. 66	69	83
Shielded part of head............................... 12¼	14	13
Snout to ear... 12	13	
Snout to anterior gular fold.......................... 12	13	
Snout to posterior gular fold 20	21	
Fore limb.. 10	16	
Hind limb 26	27	
Base of fifth to end of fourth toe..................... 9½	10	

Distribution.—Henshaw's Night Lizard has been found at Witch Creek, San Diego County, California. This' locality is in the chaparral belt, at an "altitude of about 2,700 feet." The specimen described by Prof. Cope as *X. picta* is said to have been collected in Tejon Pass, but this locality needs confirmation.

Habits.—This species lives among the granite boulders and comes out into the narrower crevices between them a few minutes before dark. It is, therefore, practicable to hunt for it only about fifteen or twenty minutes each day. If a bit of string or straw be introduced into the domain of one of these lizards it will often be seized, the reptile apparently mistaking it for some stray insect.

37.—**Xantusia riversiana** Cope. ISLAND NIGHT LIZARD.

Xantusia riversiana, COPE, Proc. Ac. Nat. Sci. Phila., 1883, p. 29 (type locality **California**); RIVERS, Am. Nat. XXIII, 1889, p. 1100 (type locality stated as **San Nicolas Island**); COPE, Proc. U. S. Nat. Mus., 1889, p. 147.

Description.—Limbs very short and body somewhat depressed. Upper surface of head very flat. Nostril pierced in a small scute at junction of rostral, internasal, postnasal, and first labial plates. Rostral broad and rather low, bounded by first labial, nasal, and internasal plates. Two internasals followed by a large hexagonal frontonasal. Behind this two prefrontals, bordered posteriorly by broad frontal and first superciliary and first supraocular plates. Each of two frontoparietal plates in contact with frontal, second, third, and fourth supraoculars, parietal, interparietal, and its fellow of opposite side. Interparietal bordered behind by two large occipitals. Latter separated from the parietals by two small scutes. A row of large supratemporals along outer edge of occipital and parietal plates. Two loreals in contact below with supralabials and above with fronto-nasal and prefrontal plates. Eye surrounded by a series of small plates, upper five of which are superciliaries. Between this ring and posterior loreal, two or three small plates. A series of four supraoculars separating super-ciliaries from frontal and frontoparietal plates. Five superior and four or five inferior labials to a point below pupil. Eye large, without lids, and with vertical pupil. Ear with a weak anterior denticulation. Inferior labials in contact with large sublabials. First pair of latter in contact on median line. Back, sides, upper and poste-rior surfaces of limbs, and gular regions, covered with

flattened granules. A series of large plates along edge of last gular fold. Quadrate ventrals in sixteen longi-tudinal and thirty-two to thirty-five transverse rows. Large preanal plates arranged in two or three series, edged by smaller scales and granules. Tail conical, covered with whorls of smooth, narrow, and transversely convex scales. A series of from ten to twelve femoral pores along each thigh.

The ground color is smoke gray or cinnamon, with numerous irregular maculations of dark brown or black. These markings are much smaller and less numerous on the lower surfaces. There is considerable variation in the color pattern. One specimen has two narrow parallel black lines, originating at the posterior edge of each occipital plate and running the whole length of the back. The space between each pair of these lines is unmarked, but the rest of the upper surface is irregularly spotted. Other specimens offer an almost perfect imitation of coarse granitic rock.

Length to anus	106
Length of tail	73*
Shielded part of head	24
Snout to ear	24
Snout to anterior gular fold	20
Snout to posterior gular fold	34
Fore limb	30
Hind limb	38
Base of fifth to end of fourth toe	14

Distribution.—This largest species of the group has been recorded from San Nicolas, Santa Catalina, and San Clemente Islands, California.

Family VIII. TEIIDÆ.

This family contains a large number of American lizards of various forms and scaling. They are most closely related to the *Lacertidæ* of the Old World. The tongue is slender and ends in two long smooth points. The head is covered with large, regular plates (except in the South American *Callopistes*). An ear-opening is

----

* Reproduced.

usually present. Eyelids are rarely wanting. Femoral and preanal pores may be either present or absent. The limbs are rudimentary in some members of the group. Two genera have been found in California.

SYNOPSIS OF GENERA.

a.—Two frontoparietal plates..............Cnemidophorus.—p. 133.
a².—One frontoparietal plate ....Verticaria.—p. 140.

Genus 18. CNEMIDOPHORUS.

"*Cnemidophorus*, Wagler, Syst. Amph., 1830, p. 154 (part)."

There are four pentadactyle limbs. The head-plates are large, except the occipitals, which are small and irregular. There are two frontoparietal plates. The back and sides are covered with small, smooth, granular scales. The ventral plates are large and arranged in both transverse and longitudinal series. The legs and tail are very long; the latter, slender and provided with large scales, which are keeled above but smooth below. The eye has well developed lids and a round pupil. Large ear-openings are present. One strong and several weaker folds cross the throat. Femoral pores are present.

SYNOPSIS OF SPECIES. *

a.—Markings on sides of head not well defined, almost obsolete; throat often suffused with slate or gray......C. tigris.—p. 134.
a².—Markings on sides of head very distinct and well defined; throat not (sometimes slightly in *C. stejnegeri*) suffused with gray or slate.

---

* Prof. Cope has described (Trans. Am. Philos. Soc., (2), XVII, 1892, p. 40, pl. IX, fig. 8) *Cnemidophorus septemvittatus* from a specimen said to have been collected by Dr. Boyle in El Dorado County, Cal., and now deposited in the United States National Museum. The registers of this museum state, in Prof. Baird's handwriting, that the number 2872, attached to this *Cnemidophorus*, belongs to two specimens of *Sceloporus*. These *Scelopori*, properly numbered, are still on the museum shelves, and it therefore seems probable that the type and only known specimen of *Cnemidophorus septemvittatus* is not of Californian origin, but has been so labeled through an erroneous duplication of museum numbers.

It may be readily distinguished from the three *Cnemidophori* known to be Californian by the presence of enlarged plates on the posterior surface of its forearm.

b.—Spots on throat few and small; central gular and collar scales
   smaller..........................C. tigris undulatus.—p. 137.
b².—Spots on throat numerous and large, often forming irregular transverse bands; central gular and collar scales larger.
                           C. stejnegeri.—p. 139.

38.—Cnemidophorus tigris Baird & Girard. DESERT WHIPTAIL.

? Ameiva tessellata, SAY, Long's Exped. Rocky Mts., 1823, II
(Philadelphia) p. 50 (London) p. 351, note 33 (type locality
Arkansas River near Castle Rock Creek").

Cnemidophorus tigris, BAIRD & GIRARD, Proc. Ac. Nat. Sci. Phila.,
VI, 1852, p, 69 (type locality **Valley of Great Salt Lake,
Utah**); BAIRD & GIRARD, Stansbury's Report Gt. Salt Lake,
p. 338, pl. II; BAIRD, U. S. Mex. Bound. Surv., 1859, II, p. 10,
pl. XXXIII; STEJNEGER, N. A. Fauna, No. 7, 1893, p. 198.

Cnemidophorus gracilis, BAIRD & GIRARD, Proc. Ac. Nat. Sci. Phila.,
VI, 1852, p. 128 (type locality **Desert of Colorado, Cal.**);
BAIRD, U. S. Mex. Bound. Surv., 1859, II, p. 10, pl. XXXIV,
figs. 7–14.

Cnemidophorus tesselatus, BAIRD, Pac. RR. Surv., X, pt. IV, p. 18.

Cnemidophorus tessellatus tessellatus, COPE, Trans. Am. Philos. Soc.
(2), XVII, pt. 1, 1892, p. 34, pl. VII.

Description.—Snout long, with nearly vertical sides.
Nostrils opening in large anterior nasal plates; latter in
contact on top of snout. Posterior nasal forming
sutures with anterior nasal, first, second and third
labials, loreal, prefrontal, and frontonasal plates.
Loreal in contact with third and fourth labials, first
subocular, preocular, first superciliary, prefrontal,
posterior nasal, and sometimes first supraocular plates.
Four supraoculars, first and fourth smaller than others.
Second, third, and fourth supraoculars separated from
superciliaries by small convex granules. Similar granules intrude between third and fourth supraoculars
and frontoparietal and parietal. Occcipitals represented
by from two to four transverse series of small plates

---

* Colorado.

behind parietals and interparietal. About five superior and six inferior labials to a point below middle of eye. Sublabials large and, except first, separated from infra-labials by small granules and plates. Anterior gulars largest centrally, becoming gradually smaller laterally and anteriorly, and changing rather abruptly to smaller posterior gulars. Central gular and collar scales a little smaller than in *C. stejnegeri*. Scales on center of collar of moderate size, those on its edge smaller. Small, smooth, convex granules on back usually slightly larger than in *C. stejnegeri*. Eight longitudinal rows of ven-tral plates. From three to six large scutes, surrounded by smaller plates and granules, in front of anus. Pos-terior surface of forearm covered with small, nearly equal-sized granules. Tail very long and provided with rings of large, obliquely keeled scales. Femoral pores varying from eighteen to twenty-three.

The color above is brownish, yellowish, or bluish gray, becoming paler toward the tail and darker on the sides, with very irregular dark and light marblings. In young specimens there are narrow light longitudinal lines sep-arated by darker bands, which are more or less broken up by spots of the same color as the lines. In older specimens these lines have become more or less obscure, and in some specimens the upper surface is nearly unicolor. The upper surfaces of the limbs are similarly colored. The dark markings on the sides of the head and neck and on the gular region are small and ill defined. The tail is gray or brown, often with dark lines along the keels of its upper scales. All the lower surfaces are creamy white, usually suffused with gray or slate on the gular region or chest, and maculated with black.

Length to anus	47	61	82	83	92	93
Length of tail	124	184	190	207	204	212
Snout to ear	11	15	19	20	20	22
Snout to interparietal	10	12	16	15	16	17
Width of head	7	9	11	12	12	14
Fore limb	17	24	29	31	32	32
Hind limb	34	47	56	56	61	65
Base of fifth to end of fourth toe	16	23	25	25	28	29

Distribution.—The Desert Whiptail Lizard or "Swift Jack" is common in many parts of the Mojave and Colorado Deserts and the Great Basin, but does not range farther west. It has been taken in Owen's, Coso, Death, Panamint, and Deep Spring Valleys, in Inyo County; at Mojave, in Kern County; Barstow, The Needles, Leach Point Valley, and Warren's Wells, in San Bernardino County; and Fort Yuma, in San Diego County, California. Its range extends across Nevada (Pahrump Valley, Oasis Valley, Pahranagat Valley, vicinity Reno) to southern Idaho (Plains near Snake River) and western Utah (Santa Clara and Great Salt Lake Valleys).

Habits.—This species lives on the open desert, over which it runs with great swiftness. The sand banks near The Needles are covered with its tracks, which end in the holes made by small mammals. So far as I have been able to learn, its food consists entirely of insects.

39.—Cnemidophorus tigris undulatus (Hallowell). CALIFORNIA WHIPTAIL.

Cnemidophorus undulatus, HALLOWELL, Proc. Ac. Nat. Sci. Phila.,
. VII, 1854, p. 94 (type locality "**Fort Yuma, San Joaquin
Valley,**" [=Fort Miller, Fresno County, Calif.]); HALLOW.,
Rept. U. S. Pac. R. R. Surv., X, pt. IV, 1859, p. 8, pl. IX, fig. 2.
Cnemidophorus tigris undulatus, STEJNEGER, N. A. Fauna, No. 7,
p. 200.

Description.—Whole animal long and slender. Nostrils opening in large anterior nasal plates; latter meeting on top of snout. Posterior nasal forming sutures with anterior nasal, first, second and third labials, loreal, prefrontal, and frontonasal plates. Loreal in contact with third and fourth labials, first subocular, preocular, first superciliary, prefrontal, posterior nasal, and first supraocular plates. Four supraoculars, first and fourth smaller than others. Second, third, and fourth supraoculars separated from superciliaries by small convex granules. Similar granules between third and fourth supraoculars and frontoparietal and parietal plates. Behind parietals, two or three transverse series of small occipitals. About five superior and six inferior labials to a point below pupils. Sublabials large, and, except first, separated from infralabials by small plates and granules. Anterior gulars largest centrally, becoming gradually smaller laterally and anteriorly, and changing abruptly to smaller posterior gulars. Largest gular and collar scales averaging smaller than in *C. stejnegeri*. Scales on center of collar moderately large, those on its edge smaller. Back covered with small, smooth, convex granules slightly larger than in *C. tigris*. Ventral plates in eight longitudinal rows. Several large plates, surrounded by smaller plates and granules, in front of anal opening. Posterior surface of forearm covered with small, nearly equal-sized granules. Long slender tail provided with rings of large, obliquely keeled scales. Femoral pores varying in number from eighteen to twenty-three.

The back is grayish or yellowish brown with about seven or nine wavy black longitudinal bands or rows of spots which are sometimes broken up into irregular marblings. On the sides of the head and neck are numerous, large, well defined black blotches. The limbs are marbled with black. The tail is yellowish or olive-brown, darkest along the keels of the upper scales. The lower surfaces are creamy or buffy white, often spotted or blotched with black; the markings on the gular region being few and usually very small.

Length to anus....................79	79	87	99	103	105
Length of tail	207	204	242	252	231
Snout to ear.......................18	18	19	23	22	24
Snout to interparietal plate...........15	15	15	18	18	19
Width of head.....................11	11	12	16	13	15
Fore limb.........................29	29	30	33	35	36
Hind limb.........................59	58	64	68	71	73·
Base of fifth to end of fourth toe......30	28	30	31	32	32

Distribution. — The California Whiptail Lizard replaces *C. tigris* in the northern, as *C. stejnegeri* does in the southern, portion of California west of the desert. Its range seems to be continuous with that of *C. tigris* through Walker and Tehachapi Passes and the Cañada de las Uvas, and thence extends north on the lower levels of the western slope of the Sierra Nevada at least as far as Mariposa County. West of the San Joaquin and Sacramento Valleys, it has been found at Los Gatos, in Santa Clara County, and at Kelseyville, in Lake County.

Habits.—Very little is known of the habits of this lizard. When hard pressed, it often tries to elude pursuit by burrowing, although it can run very swiftly. It mates, near Los Gatos, early in June.

40.—Cnemidophorus stejnegeri Van Denburgh. STEJ-NEGER'S WHIPTAIL.

Cnemidophorus stejnegeri, VAN DENBURGH, Proc. Cal. Ac. Sci. (2), IV, Pt. I, 1894, p. 300 (type locality between **San Rafael and Ensenada, Lower California, Mex.**).

Description.—Body long, with a very slender tail and very long legs. Nostrils opening in large anterior nasal plates; latter in contact on top of snout. Posterior nasal forming sutures with anterior nasal, first, second, and third labials, loreal, prefrontal, and frontonasal plates. Loreal in contact with third and fourth labials; first subocular, preocular, first superciliary, prefrontal, posterior nasal, and sometimes first supraocular plates. Four supraoculars, fourth smallest. Second, third, and fourth supraoculars separated from superciliaries by small convex granules. Similar granules between third and fourth supraoculars and frontoparietal and parietal. One to three transverse series of small occipitals behind parietals and interparietal. About five superior and five or six inferior labials to a point below pupil. Sublabials large and, except anteriorly, separated from infralabials by small granules and plates. Anterior gulars largest centrally, becoming gradually smaller laterally and anteriorly, and changing rather abruptly to smaller posterior gulars. Central gular and collar scales averaging larger than in *C. tigris* and *C. t. undulatus.* Scales on center of collar larger than those on its edge. Back covered with small, smooth, convex granules, usually slightly smaller than in *C. tigris* and *C. t. undulatus.* Ventral plates in eight longitudinal rows. From two to five large plates, surrounded by smaller plates and granules, in front of anus. Posterior surface of forearm covered with small, nearly equal-sized granules. Tail long, provided with rings of large, obliquely keeled scales.   Femoral pores varying from nineteen to twenty-five on each thigh.

The color above is yellowish or grayish brown, becoming grayer toward the head and paler on the sides, with seven or nine undulate black bands or longitudinal rows of irregular spots.   The upper surfaces of the limbs are brown or gray reticulated with black.   On the sides of the head and neck are numerous large, well defined black blotches.   The tail is yellowish or olive-brown, darkest along the keels of its upper scales.   The lower surfaces are yellowish white, rarely faintly washed with gray, usually much spotted or blotched with black; the markings on the gular region being numerous and large, often forming irregular cross-bands.

Length to anus...................... 73	89	91	93	96	98
Length of tail.....................119	229		212	247	252
Snout to ear....................... 17	21	21	20	23	23
Snout to interparietal.............. 14	17	17	17	18	18
Width of head..................... 10	13	13	12	15	15
Fore limb......................... 26	31	30	30	31	33
Hind limb......................... 53	60	58	60	63	68
Base of fifth to end of fourth toe..... 25	28	26	27	29	30

Distribution.—Stejneger's Whiptailed Lizard inhabits the western slope of the coast ranges of San Diego, Riverside, and San Bernardino Counties (Santa Ysabel Valley, Witch Creek, Julian Mountains, Clogston's Valley, San Jacinto, Hemet Valley, Lytle Creek), and probably will be found also in parts of Orange and Los Angeles Counties, California.

Habits.—Unknown, but, doubtless, similar to those of *Cnemidophorus tigris undulatus*.

Genus 19.   VERTICARIA.

Verticaria, Cope, Proc. Amer. Philos. Soc., XI, 1869, p. 158 (type hyperythra).

There are four pentadactyle limbs.   The head-plates

are large, except the occipitals, which are small and irregular. The frontoparietal plate is single. The back
and sides are covered with small, smooth granules. The
ventral plates are large and are arranged in both transverse and longitudinal series. The tail is very long and
slender and is provided with large scales, which are
keeled on its upper surface but smooth below. The eye
has well developed lids and round pupil. A large ear-
opening is present. One strong and several weaker folds
cross the throat. Long series of femoral pores are
present.

41.—Verticaria hyperythra beldingi (Stejneger). BELD-
ING'S ORANGE-THROAT.

Verticaria hyperythra, COPE, Proc. Ac. Nat. Sci. Phila., 1883, p. 32.
Verticaria beldingi, STEJNEGER, Proc. U. S. Nat. Mus., XVII, 1894,
    p. 17 (type locality Cerros Island, Lower California, Mex.).
Verticaria hyperythra beldingi, VAN DENBURGH, Proc. Cal. Ac. Sci.
    (2), V, 1895, p. 131.

Description.—Nostrils opening in large anterior nasal
plates, which meet on top of snout. Posterior nasal
forming sutures with anterior nasal, first, second, and
sometimes third labials, loreal, prefrontal, and frontonasal plates. Loreal in contact with third, fourth, and
(usually) second labials, first subocular, preocular, first
superciliary, (often) first supraocular, prefrontal, and
posterior nasal plates; sometimes divided into a larger
anterior and smaller posterior portion. Three or four
supraoculars; first in contact with superciliary, prefrontal,
and frontal plates; others separated from superciliaries
and parietal, and usually from frontoparietal, frontal,
and first supraocular, by small granular scales. A single
large frontoparietal plate separating frontal from interparietal and parietals. One or two transverse rows of
small occipital plates. About five superior and as many

inferior labials to a point below middle of eye.   Large
sublabial plates present.   Gulars large centrally, becoming smaller anteriorly and laterally, and changing abruptly to smaller granules posteriorly.   Scales on fold or
collar usually large, largest being along its edge.   Eight
longitudinal rows of ventral plates.   Back and sides
covered with small, smooth, equal-sized granules.   Limbs
plated in front and below.   Rings of large scales,
strongly keeled except on the proximal part of its ventral
surface, covering tail.   Ear-opening large, without denticulation.   About thirteen to sixteen pores in a series
along each thigh.

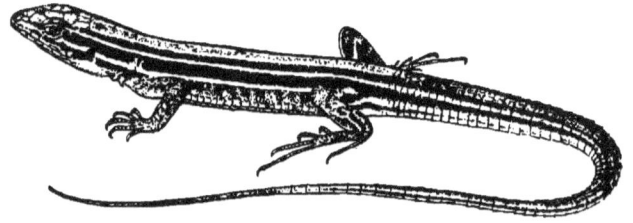

The back is black or brown, darkest in young speci
mens, sometimes dotted with gray, with three longitudinal light lines on each side.   The lower two of these
lines are wider and lighter than the upper one.   The
lowest line is continued along the side of the head and
thigh.   Near its base the tail is banded like the back,
but it becomes unicolor toward the tip.   It is bright
campanula blue in young specimens, but this color disappears with age.   The lower surfaces are yellowish
white, often tinted with gray or bluish slate on the belly,
more or less washed with bright reddish orange-chrome
in adult males.

Length to anus...... 31	36	59	61	65	68
Length of tail...... 65	77	98*	132*	166	147*
Snout to ear 8	8	13	14	14	15
Snout to interparietal plate.......... 6¼	7	10	11	11	12
Width of head 5	5	8	8	8	9
Fore limb...... 11	12	19	20	20	22
Hind limb...... 22	23	37	42	41	46
Base of fifth to end of fourth toe..... 10	11	17	19	19	20

Distribution.—Belding's Orange-throated Lizard has been found in California only in the western parts of San Diego and Riverside Counties (San Diego, Mexican border between Campo and the coast, Oak Grove, between Oceanside and San Jacinto, San Jacinto, Riverside), but ranges for some distance down the peninsula of Lower California.

Habits.—At San Jacinto this lizard lives on rocky hillsides, is very shy, and quickly retreats to holes when approached.

Family IX. SCINCIDÆ.

The tongue is slightly notched at its tip. The head is covered with large, regular plates. The scales on the body and tail are moderately large, imbricate, and reinforced with an armor of bony plates. The eyes have round pupils and well developed lids. Femoral pores are absent. Limbs may be either present or absent. An interoccipital plate is rarely present.

A single genus represents this family in California.

Genus 20. EUMECES.

Eumeces, WIEGM., Herp., Mex., 1834, p. 36 (part); *Plestiodon*, DUM. & BIBR., Erpét. Gen., V, 1839, p. 697; *Lamprosaurus*, HALLOW., Proc. Ac. Nat. Sci. Phila., 1852, p. 206 (type *guttulatus*); "*Eurylepis*, BLYTH, Journ., As. Soc. Beng., XXIII, 1854, p. 739."

The limbs are four, pentadactyle. The dorsal, lateral,

*Reproduced.

caudal, and ventral scales are thin, smooth, and strongly
imbricate. A distinct ear-opening is present. Gular
and lateral dermal folds are absent. The tail is moderately long.

SYNOPSIS OF SPECIES.

a.—Light lines persistent, the upper pair separated by two and two half
rows of scales; head never red..........E. skiltonianus.—p. 144.
a².—Light lines present in young only, upper pair separated by not more
than two rows of scales; head bright red in adults.
E. gilberti.—p. 147.

42.—Eumeces skiltonianus (Baird & Girard). WESTERN SKINK.

Plestiodon skiltonianum, BAIRD & GIRARD, Proc. Ac. Nat. Sci.
Phila., VI, 1852, p. 69 (type locality **Oregon**); BAIRD & GIRARD,
Stansbury's Exped. Gt. Salt Lake, 1853, p. 349, pl. IV, figs. 4–6.
Eumeces quadrilineatus, HALLOWELL, U. S. Pac. R. R. Surv., X, 1859,
Pt. IV, p. 10, pl. IX, fig. 3 (type locality near **Mojave River
and in San Bernardino Valley,** southern California).
Eumeces skiltonianus, COPE, Bull. U. S. Nat. Mus., No. 1, 1875, p.
45. YARROW, Bull. U. S. Nat. Mus., No. 24, 1883, p. 41; STEJ-
NEGER, N. A. Fauna, No. 7, 1893, p, 201.
Eumeces hallowellii, BOCOURT, Miss. Sci. au Mex., Rept., 6°, livr.,
1879, p. 435, pl. XXII E, fig. 7 (type locality **California**).

Description.—Body long and rounded, with long tail
and short legs. Nasal scute small, in contact with internasal, postnasal, first labial, and rostral plates. Postnasal touching nasal, internasal, anterior loreal, and
first and second labial plates. Anterior loreal forming
sutures with postnasal, internasal, (usually) frontonasal,
prefrontal, posterior loreal, and second and third labials.
Posterior loreal larger than anterior and bordered behind by two preoculars and first superciliary. Four
large supraoculars, first three touching long frontal.
Interparietal larger than either frontoparietal, but very
narrow posteriorly and sometimes not separating
parietals. Parietals very large and followed by one or
two pair of wide occipitals. Temporal plates very

large. Upper labials seven or eight in number, last largest. Symphyseal very broad and followed by one or two wide azygous sublabials (postmentals), and several large, paired sublabials in contact with infralabials. All scales on body, limbs, and tail similar in shape, very smooth, and strongly imbricate. Lower caudals of median series greatly enlarged transversely. Upper caudals about size of dorsals, larger than laterals, ventrals, and gulars. Twenty-four or twenty-six rows of scales encircling middle of body. Ear-opening about size of a gular scale and feebly denticulated anteriorly.

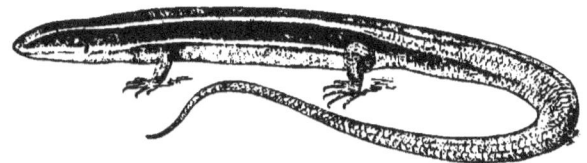

The color above is black or dark olive, with two bluish gray or pale brown lines along each side. The upper of these lines originates at the internasal plate, crosses the anterior loreal, prefrontal, supraocular, and parietal plates, and runs along the dorsal scales of the second and third rows from the median line to, and often for some distance along, the tail. The lower traverses the upper labial plates, crosses the ear-opening, and continues along the side of the neck and body to the hind limb, often reappearing on the tail. The ground color is usually darkest near the light lines. The upper pair of the latter are separated by about two and two half rows of scales. The limbs are olive, sometimes marked with darker brown on the margins of the scales. The bands of the back are continued for a varying distance on the tail, which is elsewhere greenish bluish, or grayish slate in adults, bright cobalt blue in young. The lower surfaces are yellowish white, often clouded with blue or slate on the belly and throat.

In very old specimens the ground color becomes paler and the lines widen and sometimes almost disappear.

Length to anus 29	41	55	60	64	66
Length of tail........................40	69	105	117	113	120
Snout to ear......................... 7	8	11	11	12	12
Snout to occipital plates 6	8	10	10	11	11
Fore limb........................... 7	10	15	14	16	15
Hind limb.............................10	14	22	21	23	23
Base of fifth to end of fourth toe....... 4	6	9	8	9	9

Distribution.—The Western Skink, Skilton's Skink, or Blue-tailed Lizard, is more widely distributed in California than any other saurian. It probably ranges over the entire State, except the lower, dryer portions of the Colorado and Mojave Deserts and San Joaquin Valley. It has been found in San Diego (San Diego, Cuyamaca Mountains), Riverside (San Jacinto), Los Angeles (Los Angeles), San Bernardino (Mojave River), Santa Barbara (Santa Barbara), Inyo (Argus and Panamint Mountains), Kern (Fort Tejon, Kern River), Tulare (White River, Trout Meadows), Fresno (Fresno), El Dorado, Placer (Red Point), Shasta (Pitt River, Baird), Siskiyou (Fort Jones), Lake (Kelseyville), Sonoma (Healdsburg), Napa (Napa), Marin (Larkspur), Alameda (Berkeley, Oakland), San Mateo (Pescadero), Santa Clara (Palo Alto, College Park, Mountain View, Black Mountain, Alum Rock Cañon, Smith Creek, Mt. Hamilton, Los Gatos), Santa'Cruz (Big Basin, Boulder Creek, Corralitos), and Monterey (Monterey) Counties, California.

This species lives also in western Oregon (Willamette Valley), and probably will be found in Washington.

Habits.—This lizard seems to be most abundant in

damp places, such as are found throughout the redwood forests of the Coast Range. Here it is usually found under decaying logs or behind the loose bark of old stumps. It is rather slow of movement and may easily be caught with the hands. Its food consists of insects. Vegetable matter is sometimes found in its stomach but is the food of caterpillars eaten by the lizard.

43.—Eumeces gilberti Van Denburgh. RED-HEADED SKINK.

> *Eumeces gilberti,* VAN D., Proc. Cal. Ac. Sci. (2), VI, 1896, p. 350 (type locality **Yosemite Valley, Mariposa County, California**).

Description.—Body long and rounded, with long tail and short legs. Nasal plate small, in contact with internasal, postnasal, first labial, and rostral. Postnasal touching nasal, internasal, anterior loreal, and first and second labial plates. Anterior loreal forming sutures with postnasal, internasal, frontonasal, prefrontal, posterior loreal, and second and third labials. Posterior loreal larger than anterior and bordered behind by two preoculars and first superciliary. Four large supraoculars, first three touching long frontal. Interparietal larger than either frontoparietal, but narrower than usually in *E. skiltonianus,* and often not separating parietals. Parietals very large and followed by one or two pair of wide occipitals. Temporals very large. Upper labials eight in number, eighth largest. Symphyseal very broad and followed by two wide azygous sublabials and several large, paired sublabials in contact with infralabials. All scales on body, limbs, and tail similar in shape, very smooth, and strongly imbricate. Median series of lower caudals greatly enlarged transversely. Upper caudals about size of dorsals, larger than laterals, ventrals, and gulars. Twenty-four or twenty-six rows of scales encircling middle of body. Ear-opening about size of an abdominal scale and feebly denticulate anteriorly. In old specimens of this skink, as in other species, the temporal regions become more or less swollen.

The adult is brownish olive above, slightly bronzed, or faintly washed with red, without traces of longitudinal lines. The dorsal scales are edged with darker brown, and often, especially toward the tail, show central spots of verdigris green. The tail is greenish or grayish yellow. The limbs are colored like the back. The entire head and more or less of the neck are bright poppy red slightly tinged with carmine. This color is brightest just behind the ear-opening, sometimes slightly mixed with olive on top of the head. The lower surfaces, behind the red of the throat, are dull yellowish white.

The head and back of the smallest specimen are dark seal brown, darkest on the margins of the scales, with four longitudinal light lines. The lower line on each side is indistinct, hardly to be distinguished from the coloration of the ventral surfaces, except between the ear and the fore limb. The upper pair of light lines are broader than in *E. skiltonianus* and are separated by only two rows of scales. They are white only on the head, being overlaid with bronze posteriorly. The limbs are olive, darkest on the margins of the scales. The tail is bluish gray with some bronze and greenish tints near its base. The lower surfaces are creamy white, grayish on the belly.

A somewhat larger specimen (second in table of measurements) is sepia above, with traces of the upper pair of light lines on the neck but disappearing about fifteen millimeters behind the head. The red of the

head is just beginning to appear around the ear-opening.
The lower surfaces are grayish white.

Length to anus............ 52	64	81	81	84	96
Length of tail..................... 66	119	142	136		158*
Snout to ear......:.................. 10	12	15	15	15	19
Snout to occipital plates............ 9	11	13	13	14	16
Fore limb 12	17	20	21	20	25
Hind limb........................ 18	24	29	30	30	34
Base of fifth to end of fourth toe 7	10	11	11	11	13

Distribution.—The Red-headed Skink is known only
from the western slope of the Sierra Nevada, California.†
In the vicinity of the Yosemite Valley, it has been taken
on the floor of the Yosemite Valley, at Inspiration Point,
Yosemite Valley, at an altitude of about 4,500 feet on
the Yosemite road four miles from Wawona, and between
Groveland and Crocker's. Farther north it has been
found at Big Trees, Calaveras County, and at Sugar Loaf
(5,000 feet), El Dorado County.

Habits.—This lizard is common in the mountains
near the Yosemite Valley and is well known to the
hotel keepers and ranch men. It is often seen in grass
and among rocks, retreating swiftly to holes under
stones and boulders when frightened. It seems to be
much more active than the Western Skink. Were it
not for the different position of the light stripes of the
young and the fact that this form seemingly does not
occur in most parts of the range of *E. skiltonianus*,
Eumeces gilberti might be regarded as a color phase of
the Western Skink.

---

*Reproduced.
†Dr. Stejneger's " *E. skiltonianus* " from Kern River and Fort Tejon (N. A. Fauna, No.
7, 1893, p. 202) may, perhaps, belong here.

Suborder *II.   SERPENTES—Snakes.*

Family X.   LEPTOTYPHLOPIDÆ.

There are no large plates on the belly, the body being covered everywhere with uniform scales. The head is very small and continuous with the neck. The nasal plate reaches the margin of the lip. The eye may be seen through the ocular plate. One or two large plates precede the anus. The tail ends in a small spine. A pelvic girdle is present, but there are no external traces of limbs. The lower jaw is toothed.

The small blind snakes belonging to this family are similar in appearance to the *Typhlopidæ* of the Old World and tropical America, but differ in several structural features.

Genus 21.   SIAGONODON.*

"*Siagonodon*, PETERS, Sitzb. Ges. naturf., Freunde, 1881, p. 71" (type **septemstriatus**).

The body is cylindrical, covered with smooth, cycloid scales. The rostral plate is very large, and is recurved on both the upper and lower surfaces of the protruding snout. The nasal plate is very large; behind it is the large ocular, followed in turn by wide parietal and occipital plates. A row of small scales runs along the top of the head behind the rostral plate. No supraocular plates are present. The preanal plate is not divided.

A single species represents this family in California.

44. — Siagonodon humilis (Baird & Girard).   WORM SNAKE.

Rena humilis, B. & G., Cat. N. A. Reptiles, I, Serpents, 1853, p. 143 (type locality **Valliecitas, Cal.**); STEJNEGER, Proc. U. S. Nat. Mus., XIV, 1891, p. 501; STEJNEGER, N. A. Fauna, No. 7, 1893, p. 203.

---

* The type of *Rena* is *R.* (=*Leptotyphlops*) *dulcis* B. & G.

Glauconia humilis, BOULENGER, Cat. Snakes Brit. Mus., I, 1893, p. 70.

Description.—Body long and slender, with short, blunt tail bearing a small spine at its tip. Head small, continuous with neck, slightly depressed, with prominent, rounded snout. Rostral plate strongly recurved on top of snout and continued back on lower surface of head to mouth. A large nasal plate bordering lip and divided behind, and sometimes in front of, nasal opening. Ocular plate reaching margin of lip between two labials. Two large plates, parietal and occipital, behind ocular. No supraocular plate. Nasal, ocular, parietal, and occipital plates separated from corresponding plates on opposite side of head by a single series of small, rounded, imbricate scales. Scales on chin smallest. Fourteen rows of very strongly imbricate scales around middle of body; middle ventral series often slightly enlarged. Preanal plate single. Caudal scales similar to those on body.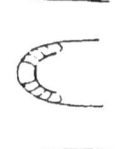

The entire upper surface, five to seven longitudinal rows of scales, is brown, sometimes slightly grayish at the edges of the scales. The lower parts are creamy white, rarely clouded with gray.

Length to anus..................91	98	133	199	235	272	291
Length of tail................. 4	4	7	10	9	9	11
Width of head................. 2	2	2	3	4	4	4
Width of middle of body....... 2	2	2½	4	5	5	6

Distribution.—In California, this little snake has been found only at Yuma and Vallecita, San Diego County, and in Death Valley, Inyo County. It probably occupies most of the intervening desert regions. Boulenger records a specimen from San Bernardino.

Family XI. BOIDÆ.

The belly is provided with a series of large plates. The head may be covered with either small scales or large plates. The eye is well developed, with vertical pupil. Rudimentary pelvis and hind limbs are present, the latter usually showing externally as a small spur on each side of the anus. Both jaws bear teeth.

Two genera of boas have been found in California.

SYNOPSIS OF GENERA.

a.—Head covered with small scales; tail not very blunt.

Lichanura.—p. 152.

a².—Head with large plates above; tail very blunt.

Charina.—p. 154.

Genus 22. LICHANURA.

Lichanura, COPE, Proc. Ac. Nat. Sci. Phila., 1861, p. 304.

The head is slightly distinct from the neck and is covered with small scales. The nostril is between two plates, the anterior of which meets that of the opposite side on the median line. The scales on the body are smooth and nearly as wide as long. The urosteges and preanal plate are undivided. The short tail is tapering, but ends in a rounded plate.

45.—Lichanura roseofusca Cope. CALIFORNIA BOA.

Lichanura roseofusca, COPE, Proc. Ac. Nat. Sci. Phila., 1868, p. 2 (type locality northern Lower California); STEJNEGER, U. S. Nat. Mus., XIV, 1891, p. 514; COPE, Proc. U. S. Nat. Mus., XIV, 1891, p. 591.

Lichanura myriolepis, COPE, Proc. Ac. Nat. Sci. Phila., 1868, p. 2 (type locality northern Lower California); STEJNEGER, Proc. U. S. Nat. Mus., XII, 1889, p. 98.

Lichanura orcutti, STEJNEGER, Proc. U. S. Nat. Mus., XII, 1889, p. 96, fig. 1 (type locality Colorado Desert, San Diego County, California); STEJNEGER, Proc. U. S. Nat. Mus., XIV, 1891, p. 513–515; COPE, Proc. U. S. Nat. Mus., XIV, 1891, p. 592.

Lichanura simplex, STEJNEGER, Proc. U. S. Nat. Mus., XII, 1889, p. 97, fig. 2 (type locality San Diego, Cal.).

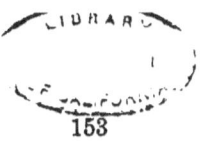

Description.—Top of head nearly flat, covered with small, smooth scales. Snout long, with a more or less prominent high rostral plate. Anterior labials very high, but their tips sometimes cut off and appearing as small scales below loreals. Loreals usually three, but their number not at all constant. Seven to ten scales encircling eye. Scales on the body smooth, imbricate, nearly as wide as long and arranged in from thirty-five to forty-three longitudinal rows, lowest row on each 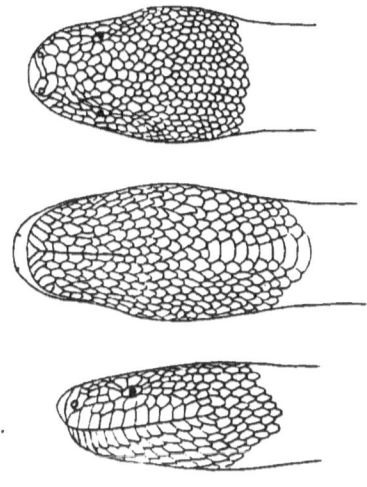 side formed of larger scales. Gastrosteges narrow and varying in number from two hundred and twenty-four to two hundred and forty-one. From thirty-nine to forty-seven urosteges. Spurs small, but easily seen at each side a little in front of anus.

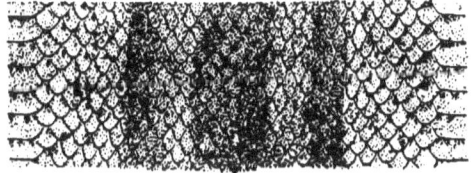

The color above is light bluish or brownish gray or deep drab, with or without three more or less indefinite reddish or yellowish brown longitudinal bands. The

middle of one of these bands originates between the
eyes, while the others arise on the temples. All or none
of these bands may extend to or along the tail. The
lower surfaces are yellowish white, more or less spotted
or blotched with brown or gray.

| Length to anus | 370 | 518* | 695* | 765* | 860* | 870* |
| Length of tail | 48 | 66 | 90 | 80 | 117 | 110 |

Distribution.—This northern relative of the great boas
of tropical America has been found, in California, only
in San Diego (San Diego, Colorado Desert, Bonsall),
Riverside (San Jacinto), San Bernardino (Cucamonga
Cañon, San Gabriel Mts.), and Los Angeles (Mt. Wilson)
Counties. It lives also in Arizona and northern Lower
California.

Genus 23. CHARINA.

"*Charina*, GRAY, Cat. Snakes Brit. Mus., 1849, p. 113 (type
bottæ);" *Wenona*, B. & G., Proc. Ac. Nat. Sci. Phila., 1852, p.
176; "*Pseudoeryx*, JAN, Arch. f. Nat., 1862, p. 242" (type
bottæ).

The head is not, or is very slightly, distinct from the
neck and is provided with large plates. The nostril is
between two plates. The scales on the body are smooth,
small, imbricate, and about as long as wide. The urosteges and preanal plate are undivided. The tail is
short, very blunt, ending in a large, rounded plate.

46.—Charina bottæ (Blainville). RUBBER SNAKE.

Tortrix bottæ, BLAINV., Nouv. Ann. Mus., IV, 1835, p. 289, pl.
XXVI, fig. 1-1b (type locality California).
Charina bottæ, GRAY, Cat. Spec. Snakes Brit. Mus., 1849, p. 113;
BOCOURT, Miss. Sci. au Mex., 8e livr., 1882, p. 511; STEJNEGER,
Proc. U. S. Nat. Mus., XIII, 1890, p. 181; COPE, Proc. U. S.
Nat. Mus., XIV, 1891, p. 592.

*From Stejneger, Proc. U. S. Nat. Mus., XIV, 1891, p. 515.

Wenona isabella, BAIRD & GIRARD, Proc. Ac. Nat. Sci. Phila., 1852,
p. 176 (type locality **Puget Sound**); GIRARD, U. S. Explor.
Exped., 1858, p. 113, Atlas, pl. VII, figs. 8-14.

Wenona plumbea, BAIRD & GIRARD, Proc. Ac. Nat. Sci. Phila., 1852,
p. 176 (type locality **Puget Sound**); B. & G., Cat. N. A. Reptiles, I, 1853, p. 139; GIRARD, U. S. Explor. Exped., Herp.,
1858, p. 112, Atlas, pl. VII, figs. 1-7; BOCOURT, Miss. Sci. au
Mex., Rept., 8e livr., 1882, p. 512, pl. XXX, figs. 7-7c.

Charina plumbea, COPE, Proc. Ac. Nat. Sci. Phila., 1861. p. 305;
TOWNSEND, Proc. U. S. Nat. Mus., X, 1887, p. 240; STEJNEGER,
Proc. U. S. Nat. Mus., XIII, 1890, p. 181; STEJNEGER, N. A.
Fauna, No. 7, 1893, p. 203.

Charina brachyops, COPE, Proc. U. S. Nat. Mus., XI, 1888, p. 88,
pl. XXXVI, figs. 2a-2f (type locality **Point Reyes, Cal.**);
STEJNEGER, Proc. U. S. Nat. Mus., XIII, 1890, p. 181; COPE,
Proc. U. S. Nat. Mus., XIV, 1891, p. 592; BOULENGER, Cat.
Snakes Brit. Mus., I, 1893, p. 131.

Description.—Top of head very slightly rounded, covered with plates which often differ greatly in size, shape,
and number in different individuals. Rostral plate very
large. Between it and broad frontal, two or three pair

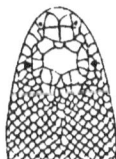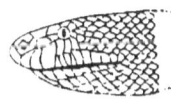

of plates—anterior nasal, internasal, and prefrontal.
Labial and prefrontal plates sometimes entering orbital
ring. ℱ A single loreal usually present,- but sometimes
two or none. Anterior upper labials usually very high,
but,˓like all head-plates, subject to much variation.*
Scales on body smooth, imbricate, about as wide as
long, and arranged in from thirty-nine to forty-nine
longitudinal rows, lowest row on each side being formed

* For an account of the scale variation in this genus, see Stejneger, Proc. U. S. Nat.
Mus., XIII, 1890, pp. 177-182.

of larger scales. Gastrosteges narrow and ranging in
number from one hundred and ninety-two to two hundred and eleven. From twenty-nine to thirty-nine
urosteges; usually all single, but sometimes a few
divided. Anal spurs small but distinct. Tail very
short and nearly as blunt as head.

All the upper and lateral surfaces are grayish, yellowish, or greenish brown, with-
out dark or light markings.
The chin and throat are sometimes clouded with gray or
brown. The rest of the lower
surface is yellowish white.

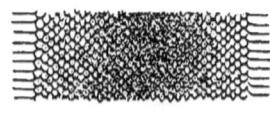

Four young, from Placer County, are dull buff both
above and below.

Length to anus	193	238	356	408	482	519
Length of tail	26	31	47	58	58	73

Distribution.—The Rubber Snake, or Two-headed
Snake as it is often called because of its blunt tail, is
not rare in the moister portions of California. It has
been taken in Siskiyou (near Mount Shasta), Lassen
(Eagle Lake), Placer (Red Point, Tahoe City), Mariposa
(Yosemite Valley), Fresno (Fresno), Tulare (Redwood
Cañon, East Fork Kaweah River), Humboldt (Humboldt
Bay), Marin (Point Reyes, Mt. Tamalpais), Alameda
(Temescal, Oakland), San Francisco (Presidio), San
Mateo (Halfmoon Bay), Santa Clara (Palo Alto, Black
Mountain), and Santa Cruz (Big Basin, Soquel)
Counties, California; and in Oregon (John Day River,
Summer Lake, Willamette Valley), Washington (Puget
Sound), Nevada (Humboldt River), and Idaho (Kootenai
County).

Habits.—This little snake is most abundant in moist
places, such as are found in the redwood forests of the

Coast Range. It is slow of movement and very gentle. When handled, it usually ties itself into a curious ball-like knot—whence its common name—but, like *Licha-nura*, never tries to defend itself by biting. A female caught in June contains large eggs.

Family XII. COLUBRIDÆ.

This family contains a large number of snakes in which the belly is covered with a series of large plates; the head-plates are large and more or less regular; the eye is always well developed, but its pupil may be either round or elliptical; there are no rudiments of limbs or pelvis; both jaws are toothed, without poison-fangs near the front of the mouth.

SYNOPSIS OF GENERA.

a.—Scales smooth.
  b.—Anal plate divided; urosteges in two series.
    c.—Loreal plate absent.
      d.—Anterior nasal united with internasal, sometimes meeting its mate on top of the snout; rostral prominent, greatly depressed; scales in thirteen rows............**Chilomeniscus**.—p. 158,
      d^2.—Internasal distinct, anterior nasal not extending onto top of head; rostral slightly prominent, not depressed; scales in fifteen rows......................**Tantilla**.—p. 176.
    c^2.—Loreal plate present.
      e.—Pupil round.
        f.—Rostral not free at edges.
          g.—Fourth infralabial largest; nasal plates usually more or less united; preoculars normally one; temporals 1-2.
            h.—Snout high, not pointed in profile.
Contia.—p. 161.
            h^2.—Snout depressed, pointed in profile.
Chionactis.—p. 159.
          g^2.—Fifth (rarely 4th or 6th) infralabial largest; nasals distinct*; preoculars normally two.
            i.—Temporals 1-1; frontal little longer than wide; a narrow white or yellow collar across nape.
Diadophis.—p. 164.

---

* Rarely united above the nostril in *Diadophis*.

i².—Temporals 2-2 (rarely 1-1 or 2-3); frontal much longer than broad; no single, definite, light collar on neck ....**Bascanion.**—p. 183.   ✗

f².—Rostral large, with free lateral edges; coloration in longitudinal bands.................**Salvadora.**—p. 180.   ✗

e².—Pupil vertically elliptical.
Rostral without free edges; coloration in blotches.
Hypsiglena.—p. 178.   ✓

b².—Anal plate single.
j.—Urosteges in one series (at least anteriorly).
Snout protruding; coloration in cross-bands.
Rhinocheilus.—p. 174.   ✓

j².—Urosteges in two series.
k.—Scales in twenty-one to twenty-three rows; loreal nearly as high as long; rostral not protruding; coloration in rings, blotches, or lines................**Lampropeltis.**—p. 166.

k².—Scales in twenty-seven to thirty-one rows; loreal elongate; rostral prominent; coloration in blotches.
Arizona.—p. 192.

a².—Dorsal scales keeled.
Laterals keeled or smooth; anal single; urosteges in two series.
l.—Prefrontals normally four (often two); scales in twenty-seven to thirty-five rows, several of the lower usually smooth; no longitudinal lines.............................**Pituophis.**—p. 195.

l²—Prefrontals two; scales in seventeen to twenty-three rows; not more than three smooth; usually with longitudinal lines.
Thamnophis.—p. 199.

Genus 24.   CHILOMENISCUS.

Chilomeniscus, Cope, Proc. Ac. Nat. Sci. Phila., 1860, p. 339 (type stramineus).

The body is stout and cylindrical, with short tail, and without constriction at neck. The snout protrudes far beyond the lower jaw and is rounded and greatly depressed. The internasal is merged in the anterior nasal which, therefore, extends onto the top of the snout. There is a small posterior nasal but no loreal. The scales are in thirteen rows, smooth and with apical pits. The anal plate is divided, and the urosteges are in two series. The eye is small, with round pupil.

47. — Chilomeniscus ephippicus Cope. BURROWING SNAKE.

Chilomeniscus ephippicus, COPE, Proc. Ac. Nat. Sci. Phila., 1867, p. 85 (type locality **Owen's Valley, California**); COUES, Surv. W. 100th Mer., V, 1875, p. 625, pl. XVIII, figs. 3, 3a.

Description.[*]—" Scales broad, in thirteen rows; tail about one-seventh total length. Rostral plate large, entirely separating internasals [anterior nasals], not encroaching on prefrontals; [posterior] nasal plate separating prefrontals and labials, in contact with preocular. Postoculars two, upper only in contact with occipital [parietal]. Superciliaries [supraoculars] very narrow, occipitals [parietals] broad as long. Temporals $\frac{1}{1}$ [1+1] large. Labials above, seven, third and fourth in orbit, these with second, narrow erect; first longitudinal; fifth and sixth smaller than the others, seventh suddenly larger. Inferior labials eight, first pair in contact before pregenials; postgenials very small. Gastrosteges 113, separated from geneials by four rows gulars; anal 1–1; urosteges 28–28.

"Above reddish or yellowish, with twenty-one black cross-bars to vent, which are broader than interspaces, and do not quite reach gastrosteges; five nearly complete rings on tail. Belly white. From occipitals [parietals] to anterior part frontal with the labials opposite this part (except their lower edges) black.

" Total length five and one-half inches."

Distribution.—The only Californian locality at which this snake has been taken is Owen's Valley, Inyo County. It has been found in Arizona.

Genus 25. CHIONACTIS.

Chionactis, COPE, Proc. Ac. Nat. Sci. Phil., 1860, p. 241 (type occipitale).

The body is small but not very slender, with short,

---

[*] Original description by Cope.

tapering tail, and little if any constriction at neck. The snout is long, rounded, and much depressed. The head-plates are normal except in the union of the anterior and posterior nasals. One preocular, two postoculars, and a loreal are present. Temporals are normally 1–2. The scales are smooth, in fifteen rows. The anal plate is divided and the urosteges are in two series. The eye is rather small, with round pupil.

48.—Chionactis occipitalis (Hallowell). DESERT SNAKE.

Rhinostoma occipitale, HALLOW., Proc. Ac. Nat. Sci. Phil., VII, 1854, p. 95 (type locality **Mojave Desert**).

Lamprosoma occipitale, HALLOW., l. c., VIII, 1856, p. 310; HALLOW., U. S. Pac. R. R. Surv., X, 1859, pt. IV, p. 15, pl. IV, figs. 2a–2c; KENN., U. S. Mex. Bound. Surv., III, Rept., 1859, p. 21, pl. XXI, fig. 1; BOCOURT, Miss. Sci. au Mex., Rept., 9e Livr., 1883, p. 558, pl. XXXIV, fig. 6–6e.

Lamprosoma annulatum, BAIRD, U. S. Mex. Bound. Surv., III, Rept., 1859, p. 22, pl. XXI, fig. 1 (type locality **Colorado Desert**).

Chionactis occipitalis, COPE, Proc. Ac. Nat. Sci. Phil., 1866, p. 310; COPE, Proc. U. S. Nat. Mus., XIV, 1891, p. 605.

Contia occipitalis, GARMAN, Mem. Mus. Compr. Zool., VIII, 3, 1883, p. 91; BOULENGER, Cat. Snakes Brit. Mus., II, 1894, p. 266.

Description.—"Head small, of same breadth posteriorly as neck, depressed in front; snout rounded; rostral

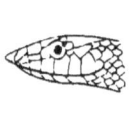 plate large, excavated below, presenting a triangular shape above and in front where it forms the extremity, of the muzzle ; internasals smaller than prefrontals, their inner margins much shorter than their external, which are in contact with the upper margins of the nasal plates; the prefrontals are more or less pentangular in shape, the posterior margin of each in contact with the anterior margin of the antocular, the supraocular, and

the half of the frontal plate, its external margin with the upper margin of the frenal; the frontal plate is about as broad as long, narrower posteriorly; supraoculars broader posteriorly; occipitals of moderate size, pentangular; nostril large, deeply excavated, in nearly the center of a large and conspicuous nasal plate, somewhat pyriform; a long and very narrow frenal, lying between the second and third supralabials, and the prefrontal; but one preocular, which is quadrangular, resting on the third supralabial; two postoculars, the upper much larger than the lower; there are seven supralabials, the three anterior smaller considerably than those which follow; the eye in contact inferiorly with the third and fourth; body long and slender, depressed; scales, of which there are fifteen rows, quadrangular, smooth and shining, their posterior margins rounded, the three inferior rows larger than the others; gastrostiga appearing to a slight extent upon the flanks; tail short, with a somewhat blunt extremity." Gastrosteges 147–158. Urosteges 34–44.

"Milk white above, with thirty-four transverse black bands, including one upon the posterior part of the head; six complete rings of black upon the tail, and one incomplete just behind the anus; jaws, chin, throat and abdomen white; interspaces between rings upon under part of tail white."

Length to anus 229
Length of tail.. 56

Distribution.—Mojave and Colorado Deserts.

Genus 26. CONTIA.

Contia, B. & G., Cat. N. A. Rept., I, Serp., 1853, p. 110 (type **mitis**);
Lodia, B. & G., l. c., p. 116 (type **tenuis**); ?*Sonora,* B. & G.,
l. c., p. 117 (type **semiannulata**); "*Eirenis,* Jan, Elenco Sist.
d. Ofidi, 1863, p. 48."

The body is rather stout for so small a snake, with short, tapering tail, and slight constriction at neck. The head is flat-topped, with broad, rounded snout. Its plates are normal, except that the anterior and posterior nasals usually unite above, or both above and below, the nostril. Usually one preocular and two postoculars are present. Temporals are 1–2. There is one loreal. The scales are smooth, in fifteen or seventeen rows, each with one apical pit. The anal plate is divided, and the urosteges are in two series. The eye is small, with round pupil.

49.—Contia mitis Baird & Girard. SHARP-TAILED SNAKE.

Contia mitis, B. & G., Cat. N. A. Rept., I, Serpents, 1853, p. 110 (type locality **San José, Calif.**); YARROW, Bull. U. S. Nat. Mus., No. 24, 1883, p. 87; BOULENGER, Cat. Snakes Brit. Mus., II, 1894, p. 267.

? Lodia tenuis, B. & G., Cat. N. A. Rept., Pt. I, Serp., 1853, p. 116 (type locality **Puget Sound, Or.**); GIRARD, U. S. Explor. Exped., Herp., 1858, p. 122, pl. IX, figs. 8–11; COPE, Proc. U. S. Nat. Mus., XIV, 1891, p. 602.

Ablabes purpureocauda, GUNTHER, Cat. Colub. Snakes, Brit. Mus., 1858, p. 245 (type locality **California**).

Description.—Head wide, with flattened top and broad, rounded snout. Rostral plate large, high, hollowed below, and bounded behind by internasal, nasal, and first labial plates. Plates on top of head, a pair of in-

  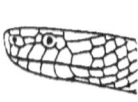

ternasals, a pair of prefrontals, a frontal broad in front but pointed behind, a long supraocular on each side, and a pair of large parietals. Anterior and posterior nasal plates frequently united above, or both above and below, nostril. Loreal small and nearly square. Normally one preocular, but two sometimes present. Postoculars two,

rarely one. Temporals one followed by two. Seven
superior and seven inferior labials, sixth upper and
fourth lower largest, third and fourth upper bordering
eye, first pair of lower meeting on median line behind
small triangular mental. Genieals in two pair, anterior
much larger than posterior. Scales on body smooth, in
fifteen rows. Anal plate divided. Gastrosteges varying in number from one hundred and fifty-one to one
hundred and eighty-six. Urosteges in two series of from
twenty-nine to fifty-two. Tail short, conical, ending in
a sharply pointed plate.

The color above is grayish or yellowish brown, usually
very finely punctulated or reticulated with slate or black,
with or without a light yellowish or brownish line along
each side. The scales below these lines are sometimes
spotted with black. In very young specimens a continuous black line along each side takes the
place of these spots, while a similar line
runs along the middle of the back. The
sides of the head show these lateral black lines more
or less distinctly. The tail is colored like the back,
except that its upper surface is sometimes suffused with
red. The lower surfaces are grayish or yellowish white,
transversely barred with black on the anterior half of
each gastrostege and (often) urostege.

| Length to anus | 106 | 176 | 238 | 240 | 299 | 330 |
| Length of tail | 22 | 27 | 57 | 37 | 78 | 83 |

Distribution.— This harmless little snake is common
among the redwoods of Santa Cruz (Big Basin) and San
Mateo (Woodside) Counties, California. It has been
found also in Santa Clara (San José), Alameda (Haywards, Alameda), Marin, Sonoma (Petaluma), Mendocino (Eel River Bridge), and Shasta Counties. Dr.
Yarrow has recorded it from Fresno. I have seen a

specimen from Fyffe, El Dorado County. It has not
been taken in the southern portion of California, but
ranges north across Oregon and Washington to Puget
Sound.

Habits.—Unknown.

Genus 27. DIADOPHIS.

Diadophis, B. & G., Cat. N. A. Rept., I, Serp., 1853, p. 112 (type
punctatus).

The body is slender, with long, tapering tail, and
slight constriction at neck. The head is flat-topped,
with broad, rounded snout. Its plates are normal.
The nasal plates very rarely unite above the nostril.
There are two preoculars and two postoculars. Temporals are 1-1. A loreal is present. The scales are
smooth, in fifteen or seventeen rows, each with one apical pit. The anal plate is divided, and the urosteges
are in two series. The eye is moderately large, with
round pupil.

50.— **Diadophis amabilis** Baird & Girard. WESTERN
RING-NECK SNAKE.

Diadophis amabilis, BAIRD & GIRARD, Cat. N. A. Rept., Pt. I, Serpents, 1853, p. 113 (type locality San José, Cal.); BAIRD, U. S.
Pac. R. R. Surv., X, 1859, Pt. III, pl. XXXIII, figs. 83.

Diadophis pulchellus, B. & G., Cat. N. A. Rept., Pt. I, Serp., 1853,
p. 115 (type locality El Dorado Co., Cal.); BAIRD, U. S. Pac.
R. R. Surv., X, 1859, Pt. III, pl. XXXIII, figs. 85; STEJNEGER,
N. A. Fauna, No. 7, 1893, p. 203.

Diadophis punctatus pulchellus, COPE, Proc. Ac. Nat. Sci. Phila.,
1883, p. 27.

Diadophis punctatus amabilis, YARROW, Bull. U. S. Nat. Mus., 24,
1882, p. 95 (part); TOWNSEND, Proc. U. S. Nat. Mus., 1887, p.
239.

Diadophis amabilis pulchellus, COPE, Proc. U. S. Nat. Mus., XIV,
1891 (1892), p. 616.

Diadophis amabilis amabilis, COPE, Proc. U. S. Nat. Mus., XIV,
1891 (1892), p. 616.

Coronella amabilis, BOULENGER, Cat. Snakes Brit. Mus., II, 1894, p.
207 (part).

Description.— Top of head flattened posteriorly, but curving slightly downward to broad, rounded snout. Rostral plate large, broader than high, hollowed below, and bounded behind by internasal, anterior nasal, and first labial plates. Plates on top of head, a pair of internasals, a pair of prefrontals, a short, broad frontal between two supraoculars, and a pair of long parietals. Anterior and posterior nasals normally distinct, but rarely united above nostril. Loreal small and nearly square. Preoculars and postoculars two each. Temporals one followed by one.

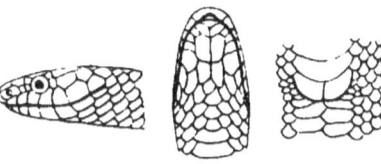

Seven (rarely eight) superior and eight (rarely seven) inferior labials, sixth (or seventh) upper and fifth (or fourth) lower largest, third and fourth (rarely fourth and fifth) superior reaching eye, first pair of inferior meeting on median line. Geneials in two pair, anterior very slightly, if at all, larger than posterior pair. Scales on body smooth, in fifteen rows. Anal plate divided. Gastrosteges varying in number from one hundred and eighty-two to two hundred and ten. Urosteges in two series of from fifty-three to seventy. Tail tapering, ending in a pointed, conical plate.

The color above is olive, brownish, greenish, bluish, or blackish slate, or gray, minutely reticulated, but without definite markings, except a light collar across the neck just behind the head. This collar may be white, yellow, or unicolor with the belly, and is often edged with black or slate. It covers from one and one-half to three transverse rows of scales. The upper part of the head is usually a darker shade of

the color of the back.   The upper labials are partly
white or yellow.   The lower surfaces, including none,
one-half, one, one and one-half, or two rows of scales,
are white, yellow, rose,* orange, lake,* or coral red—
brightest posteriorly—more or less spotted with black
at the posterior edges of the gastrosteges and urosteges
and under the head.

Length to anus.................137   166   270   333   366   420
Length of tail................. 32   40   68   66   101   92

Distribution.— The Western Ring-neck or "Red-
bellied" Snake lives in all parts of California except
the desert area.   It is rather inconspicuous, because of
its small size, but has been taken in San Diego (San
Diego, vic. Carlsbad and Oceanside), San Bernardino
(San Bernardino, Ontario), Fresno (Fresno), Mariposa
(Mariposa, Yosemite Valley), El Dorado, Santa Cruz
(Santa Cruz), Santa Clara (San José, Palo Alto), Alameda (Oakland), Contra Costa (Mount Diablo), Marin,
Sonoma (Sonoma, Petaluma, Healdsburg), Napa (Calistoga), Lake (Highland Springs), and Shasta (McCloud
River) Counties, California, and in Willamette Valley
and at Fort Dalles, Oregon.

Habits.—*Diadophis amabilis* is most often found under
boards or logs in moist localities, sometimes even in salt
marshes.   Its food probably consists chiefly of insects,
but one specimen had eaten a half-grown tree-toad
(Hyla regilla).   Nothing is known of its breeding habits.

Genus 28.   LAMPROPELTIS.

Lampropeltis, FITZINGER, Syst. Rept., 1843, p. 25 (type **getulus**);
  Sphenophis, FITZ., l. c., p. 25 (type **coccinea**); *Ophibolus,* B. &
  G., Cat. N. A. Serp., 1853. p. 82 (type **sayi**); *Bellophis,* LOCK-
  INGTON, Proc. Cal. Acad. Sci., VII, 1877, p. 52 (type **zonatus**).

The body is rather thick, with short tail, and little if

*In formalin.

any constriction at neck. The snout is broad and high. The upper head-plates are normal. The nasal plates are distinct. One (rarely two) preocular and two (rarely one or three) postoculars are present, as is also a small loreal plate. Temporals are normally 2–3, rarely 1–2, 1–3, 2–2, 2–4, ór 3–4. The scales are smooth, in twenty-one or twenty-three (or 24) rows, each with two apical pits. The anal plate is undivided, but the urosteges are in two series. The eye is of moderate size, with round pupil.

SYNOPSIS OF SPECIES.

a.—Black rings more or less split by red; gastrosteges fewer than 220.

L. zonata.—p. 167.

a².—No red; gastrosteges more than 220.

b.—Color in transverse blotches or rings.....L. boylii.—p. 169.

b².—Color chiefly in longitudinal lines or blotches.

L. californiæ.—p. 172.

51. — Lampropeltis zonata (Blainville). CALIFORNIA KING SNAKE.

Coluber (Zacholus) zonatus, BLAIN., Nouv. Ann. du Mus., IV, 1835, p. 293 (type locality California); B. & G., Cat. N. A. Rept., Pt. I, Serp., 1853, p. 153.

Ophibolus pyrrhomelas, COPE, Bull. U. S. Nat. Mus., I, 1875, p. 37 (part?); COPE, Proc. U. S. Nat. Mus., XIV, 1891, p. 610 (part?).

Bellophis zonatus, LOCKINGTON, Proc. Cal. Acad. Sci., VII, 1877, p. 52 (type locality [Santa Barbara,] Calif.).

Ophibolus getulus multicinctus, YARROW, Proc. U. S. Nat. Mus., V, 1882, p. 440 (type locality Fresno, California).

Coronella multifasciata, BOCOURT, Miss. Sci. au Mex., Rept. 10e Livr., 1886, p. 616, pl. XL, figs. 2–2c (type locality California).

Coronella zonata, BOULENGER, Cat. Snakes Brit. Mus., II, 1894, p. 202.

Description.—Top of head slightly flattened posteriorly, curving downward to broad rounded snout. Temporal regions frequently swollen. Rostral plate large, broader than high, hollowed below, and bounded behind by internasal, anterior nasal, and first labial plates. Plates on top of head, a pair of internasals, a pair of prefrontals,

a short, broad, irregularly wedge-shaped frontal between
two supraoculars, and a pair of large parietals. Anterior
and posterior nasals distinct. A small loreal present,
but sometimes united with prefrontal. Two postoculars
and one (or rarely two) preocular. Temporals normally
two followed by three, sometimes 1–2, 1–3, or 2–4. Seven
(rarely six) superior and nine inferior labials, fifth and
sixth superior and fifth (or fourth) inferior largest, third
and fourth superior reaching eye, first pair of inferior
meeting on median line. Geneials in two pair, anterior
larger than posterior. Scales on body smooth, thin,
imbricate, in twenty-one or twenty-three rows. Anal
plate never divided. Gastrosteges varying in number
from one hundred and ninety-nine to two hundred and
fifteen. Urosteges in two series of from forty-five to
sixty-one. Tail short but slender.

The snout may be black, white, or spotted. The middle third of the head is black. A white band crosses
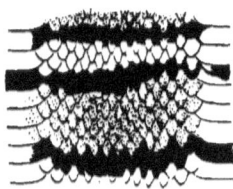 the back of the head, involving
the tips of the parietal plates and
joining the white of the throat.
Behind this white one is a full or
half ring of black, followed in
turn by another of red. The
whole body is similarly marked,
being encircled by from twenty-five to forty-three white
rings* between which are rings of black more or less
divided and replaced by blotches or rings of red or pink.
The proportion of black to red varies greatly in different
specimens, as does also the intensity of the red. This
color is sometimes present anteriorly only, and is usually
absent near the tip of the tail. The colors of the back

* Not counting the five to eleven on the tail.

and sides are continued, somewhat irregularly, onto the lower surfaces. The white areas, and more rarely the red ones also, are sometimes tinged with dull yellowish brown. The white rings are little if at all broader on the sides than on the back.

Length to anus	288	486	560	607	695	722
Length of tail	46	71	97	111	118	124

Distribution.—This brilliant snake seems to prefer the moister, cooler portions of the State, such as are occupied by coniferous forests. It has been taken in San Diego (vic. San Diego), San Bernardino (San Bernardino Mountains), Santa Barbara (Santa Barbara), Tulare (Heaven's Gate, near Little Kern Lake), Fresno, Tuolumne (Hodgdon's), Mariposa (Yosemite Valley), El Dorado (Riverton), Santa Cruz (Soquel, Santa Cruz, Glenwood), and Santa Clara (Mt. Hamilton) Counties, California.

Habits.—Very little is known of the habits of this snake. Old hunters say that it destroys many rattlers and other snakes. One of my specimens had eaten two Blue-bellied Lizards (*Sceloporus occidentalis*).

Many names are applied to this species, among which are King Snake, Red Milk Snake, Coral or Corral Snake, Ring Snake, Harlequin Snake, etc. It is popularly supposed to be very poisonous, but, like all Californian reptiles excepting the rattlesnakes, is entirely harmless.

52.—Lampropeltis boylii Baird & Girard. BOYLE'S MILK SNAKE.

Ophibolus Boylii, B. & G., Cat. N. A. Rept., Pt. I, Serp. 1853, p. 82 (type locality **El Dorado Co., Cal.**).

Coronella balteata, HALLOW., Proc. Ac. Nat. Sci. Phila., VI, 1853, p. 236; *Id.*, Rep. Pac. R. R. Surv., X, pt. 4, 1857, pp. 14, 24, pl. V (type locality "**El Paso Creek and Benicia**; also intermediate places," **California**).

Lampropeltis boylii, COPE, Proc. Ac. Nat. Sci. Phila., 1860, p. 255; STEJNEGER, N. A. Fauna, No. 7, 1893, p. 204.

"*Coronella getulus,* var. *pseudogetulus,* JAN, Icon. Gen. Ofid., I, Livr. 12, pl. VI, Fig. 2."

Ophibolus getulus boylii, COPE, Proc. U.S. Nat. Mus., XIV. 1891, p. 613.

Coronella getula, BOULENGER, Cat. Snakes Brit. Mus., II, 1894, p. 197 (part).

Description.—A larger and stouter snake than *L. zonata.* Top of head slightly flattened posteriorly, curving downward to broad rounded snout. Temporal regions rarely if ever swollen. Rostral plate large, little broader than high, hollowed below and bounded behind by internasal, anterior nasal, and first labial plates. Plates on top of head, a pair 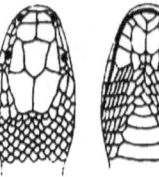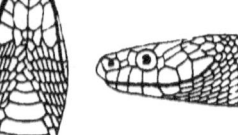 of internasals, a pair of prefrontals, a short, broad, irregularly wedge-shaped frontal, supraocular of each side, and a pair of large parietals. Anterior and posterior nasals distinct. A small loreal present, but sometimes united with posterior nasal. One preocular and two (rarely one) postoculars. Temporals normally two followed by three, but may be 2-2, 2-4, or 3-4. Seven superior and eight to ten inferior labials, fifth and sixth superior and fifth or fourth inferior largest, third and fourth superior reaching eye, first pair of inferior meeting on median line. Geneials in two pair, anterior much larger than posterior. Scales smooth, thin, imbricate, in twenty-three rows. Anal plate undivided. Gastrosteges varying in number from two hundred and twenty-one to two hundred and fifty-five. Urosteges in two series of from forty-six to sixty, a few of the first sometimes undivided.

The snout and sides of the head are yellow or white,

more or less spotted or blotched with dark brown along
the edges of the plates. The nape and the top of the

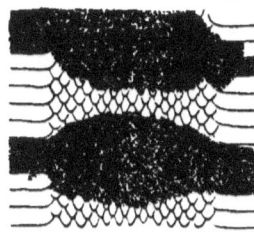 head behind the prefrontal
plates are dark brown, with a
varying number of white or
yellow spots, one of which is
very constantly present just
behind the parietal plates.
The body and tail are marked
with great blotches or rings
of brown, separated by nar-
rower rings of yellow or white. These white rings are
much broader on the sides than near the middle of the
back, and vary in number from twenty-four to thirty-
five on the body and five to eight on the tail. The
markings of the sides are continued onto the lower
surfaces.

Length to anus	317	383	586	733	921	954
Length of tail	44	55	93	129	118	135

Distribution.—Boyle's Milk Snake is common in al-
most all parts of California, where it has been taken in
San Diego (San Diego, Santa Margarita), Riverside (San
Jacinto, Riverside), Los Angeles (Pasadena), Kern (Ft.
Tejon, Kern Valley), Tulare (East Fork Kaweah River,
Three Rivers), Fresno (Fresno), El Dorado (Alt. 2000
ft.), Placer (Applegate), Santa Barbara (Santa Barbara),
Santa Clara (Los Gatos, Palo Alto), Alameda (Oakland),
Solano (Benicia), San Francisco, Marin (Mt. Tamal-
pais, Camp Taylor), Sonoma (Healdsburg), Mendocino
(Irishes), and Shasta (Redding, Ft. Reading, McCloud
River) Counties.

It has been recorded from St. Thomas and Overton,
Muddy Valley, Nevada.

Habits.—The black and white king snake is most

abundant where the country is covered with chaparral
and where small streams are numerous.   It is usually
very gentle, but sometimes fights its captor most fiercely, rarely, however, being able to draw blood with its
small teeth.   I have twice found it swallowing the contents of quail's nests, and once observed one crawling
along the ground and looking up into the bushes for
nests of small birds.   Several times while I watched its
quick eyes detected nests three or four feet above it, but
although the snake immediately climbed up to these, it
did not obtain a meal, for the nests which it examined
had been abandoned by their builders or robbed by some
earlier comer.

While I was watching a man spade up a small plot of
ground, he killed two gophers *(Thomomys)* and threw
them a few feet away.   A few minutes later a snake of
this species appeared, went directly to the spot where
the gophers lay side by side, and swallowed first the
adult and then the half grown one.   It took no notice
of our presence, and after completing its hearty meal
disappeared in the direction whence it had come.

53.—Lampropeltis californiæ (Blainville).   CALIFORNIA
MILK SNAKE.

Coluber (Ophis) Californiæ, BLAINV., Nouv. Ann. du Mus., IV, 1835,
    p. 292, pl. XXVII, figs. 1-1b (type locality **California**); B. &
    G., Cat. N. A. Rept., Pt. I, Serp., 1853, p. 153.
Coronella californiæ, DUM. & BIBR., Erp. Gen., VII, 1854, p. 623.
Ophibolus californiæ, COPE, Bull. U. S. Nat. Mus., No. 1, 1875, p. 37.
Ophibolus getulus eiseni, YARROW, Proc. U. S. Nat. Mus., V, 1882,
    p. 439 (type locality **Fresno, California**).
Ophibolus getulus californiæ, COPE, Proc. U. S. Nat. Mus., XIV, 1891,
    p. 614.
Coronella getula, BOULENGER, Cat. Snakes Brit. Mus., II, 1894, p.
    197 (part).

Description.—Similar to *L. boylii* in everything but
color.   Top of head is slightly flattened posteriorly,

curving downward to broad rounded snout. Temporal regions rarely if ever swollen. Rostral plate large, little broader than high, hollowed below, and bounded behind by internasal, anterior nasal, and first labial plates. Plates on top of head, a pair of internasals, a pair of prefrontals, a short, broad, irregularly wedge-shaped frontal, supraocular of each side, and a pair of large parietals. Anterior and posterior nasals distinct. A small loreal. One preocular and two (rarely one or three) postoculars. Temporals normally two followed by three. Seven superior and nine or ten inferior labials, fifth and sixth superior and fifth inferior largest, third and fourth superior reaching eye, first pair of inferior meeting on median line. Geneials in two pair, anterior much larger than posterior. Scales smooth, thin, imbricate, in twenty-three (or twenty-four) rows. Anal plate undivided. Gastrosteges varying in number from two hundred and twenty-six to two hundred and thirty-six. Urosteges in two series of from fifty to fifty-eight.

This is a very peculiar snake, which may prove to be a mere variation of *Lampropeltis boylii*, from which it does not differ in size, form, or scale characters. There is an immense amount of variation in the color pattern; indeed this is rarely alike in any two specimens. The head is not colored differently from that of *L. boylii*, except that there often is more yellow near the posterior edges of the parietal plates. Along the sides of the body are more or less broken longitudinal lines or bands 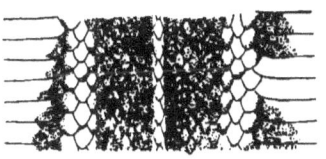 of white or yellow. Above these the coloration is dark brown to the median line, along which is a single, definite, narrow line, or a series of small spots or blotches, or both. The tail is dark brown, spotted above with white or yellow. The gastrosteges are yellow or white, unicolor or blotched with brown as in *L. boylii.*

Length to anus...277   342
Length of tail..44   49

Distribution.—The California Milk Snake has been found in San Diego, Riverside, and Fresno Counties.

Habits.—Unknown, but probably like those of *L. boylii.*

Genus 29. RHINOCHEILUS.

Rhinocheilus, B. & G., Cat. N. A. Rept., I, Serp., 1853, p. 120 (type lecontei).

The body is rather slender, with short, tapering tail. The head is slightly distinct from the neck, and ends in a narrow snout which projects far beyond the lower jaw. The head-plates are normal. The nasal plates rarely unite above the nostril. One (or two) preoculars and two (or three) postoculars are present, as is also a small loreal. Temporals are normally two followed by three. The scales are smooth, in seventeen to twenty-five rows, with apical pits. The anal plate is divided. Urosteges are in one series, except usually toward the end of the tail. The eye is of moderate size, with round pupil.

54.—Rhinocheilus lecontei Baird & Girard. LONG-NOSED SNAKE.

Rhinocheilus Lecontei, B. & G., Cat. N. A. Rept., Pt. I, Serp., 1853, p. 120 (type locality **San Diego, California**); BAIRD, U. S. Pac. R. R. Surv., X, 1859, Pt. III, pl. XXIII, figs. 90; BAIRD, U. S. Mex. Bound. Surv., Rept., 1859, pl. XX; BOCOURT, Miss. Sci. au Mex., 1886, p. 602, pl. XL, figs. 7-7d; COPE, Proc. U. S. Nat. Mus., XIV, 1891, p. 606; BOULENGER, Cat. Snakes, Brit. Mus., II, 1894, p. 212.

Description.—Head round and snout projecting and

pointed. Temporal regions not swollen. Rostral plate large, prominent, recurved on top of snout, and bounded behind by internasal, anterior nasal, and first labial plates. Plates on top of head, a pair of internasals, a pair of prefrontals, a broad frontal, supraocular of each side, and a pair of rather short, rounded parietals. Anterior and posterior nasals usually distinct, but sometimes united above nostril. Loreal small, elongate, sometimes entering orbit. One or two pre-   and two or three postoculars. Temporals normally two followed by three, rarely 2-2, 2-3, or 2-4. Eight (rarely nine) superior and nine (or ten) inferior labials, seventh or eighth superior and fifth or sixth inferior largest, fourth and fifth (or fifth and sixth) superior reaching eye, first pair of inferior meeting on median line. One or two pair of geneials, posterior very narrow when present. Scales on body smooth, thin, in twenty-three or twenty-five rows. Anal plate not divided. Gastrosteges varying in number from one hundred and ninety-one to two hundred and six. Urosteges in one series, or more often in one series anteriorly and two posteriorly, of from forty to fifty-five.

The snout is yellowish more or less marked with black. Back of this the head is black or brown, often spotted with yellow or white. Across the back is a series of large black or brown blotches; twenty to twenty-eight on the body and six to eight on the tail. These blotches may be rounded, pointed, or truncate on the sides, and are from one and one-half to three times as long as the intervals which separate them.

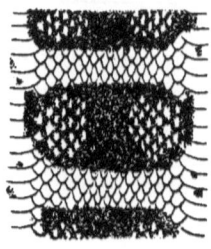 These intervals are brick red, orange, yellow, or white, and usually are continuous with the white or yellow of the lower surfaces. The lateral scales which fall within the dark blotches often have light centers, while those in the light intervals are similarly spotted with black or brown. Small black or brown blotches are sometimes present on the sides midway between the larger ones. The lower surfaces are yellow or white, unicolor or marbled with black or brown.

| Length to anus | 275 | 310 | 490 | 496 | 520 | 800 |
| Length of tail | 43 | 51 | 70 | 73 | 83 | 140 |

Distribution.—It is probable that this snake occurs throughout southern California. It has been taken at San Diego, San Diego County, and White River, Tulare County. It ranges east across Arizona.

Habits.—Unknown.

Genus 30.   TANTILLA.

Tantilla, B. & G., Cat. N. A. Rept., I, Serp., 1853, p. 131 (type coronata); *Homalocranion*, Duméril, Mém. Ac. Sci. Paris, XXIII, 1853, p. 490.

The body is very slender, with tail of moderate length and no constriction at neck. The snout protrudes a little beyond the lower jaw. The head is very low, and very flat above. Its plates are normal, except that there is no loreal. One preocular and one or two postoculars are present. The scales are smooth, arranged in fifteen (or thirteen) rows. The anal plate is divided, and the urosteges are in two series. The eye is small, with round pupil

55. Tantilla eiseni Stejneger. CALIFORNIA TANTILLA.

Tantilla nigriceps, YARROW, Bull. U. S. Nat. Mus., No. 24, 1883, p. 85 (part).

Tantilla eiseni, STEJNEGER, Proc. U. S. Nat. Mus., XVIII, 1896, p. 117 (type locality **Fresno, California**).

Description.*—"Head very flat above, rather broad across the anterior temporals; eyes small; rostral wider than high, the portion visible from above longer than the internasal suture; internasals short; prefrontals nearly twice as large as internasals, their lower border wedged in between posterior nasal and preocular, but not in contact with supralabials; frontal rather long, six-sided, angular in front and behind, the lateral borders nearly parallel; supraoculars rather small, half as wide as frontal; parietals long and narrow, nearly as long as their distance from tip of snout; nasals long, the posterior in contact with preocular, which is but slightly shorter; no loreal; one preocular; two postoculars; temporals long, 1+1; supralabials seven, last one largest, third and fourth entering eye; sublabials [infralabials] seven, four in contact with first pair of chin shields; first pair of sublabials not in contact behind mental; fifteen rows of smooth scales; four rows of scales between posterior chin shields and ventrals; ventrals 176 [167–181]; anal divided; subcaudals 62+1 [58–65].

"Color (in alcohol) uniform pale flesh color, slightly darker grayish brown above; top of head, lores, temples, and nape for a distance of three scale-lengths back of the parietals, dark grayish-brown; behind this a narrow white [transverse] band, one scale-length wide, bordered behind by a few dark-brown dots.

"Total length, 365; tail, 82 mm."

*Original description of Stejneger.

Distribution.— Seven specimens of this little snake were collected by Dr. Gustav Eisen near Fresno, Fresno County, California, in 1879. The species has not been found since.

Genus 31. HYPSIGLENA.

Hypsiglena, Cope, Proc. Ac. Nat. Sci. Phila., 1860, p. 246 (type ochrorhynchus); *Pseudodipsas*, Peters, Mon. Berl. Ac., 1860, p. 521; *Comastes*, Jan, Elenco Sist. Ofid., 1863, p. 102.

The body is small, with moderate, slender tail. The head is distinct from the neck by reason of the swollen temples, which in old specimens are greatly enlarged. The snout is rounded and rather prominent. The head plates are normal. The nasals rarely unite above the nostril. Two (or three) preoculars and two postoculars are present, as is also a loreal. Temporals are normally one followed by two. The scales are smooth, in nineteen or twenty-one rows, with apical pits. The anal plate is divided. Urosteges are in two rows. The eye is of moderate size or small, with vertically elliptic pupil.

56.—Hypsiglena ochrorhynchus Cope. Spotted Night Snake.

Hypsiglena ochrorhynchus, Cope, Proc. Ac. Nat. Sci. Phila., 1860, p. 246 (type locality **Cape San Lucas, Lower California, Mex.**); Cope, Proc. Ac. Nat. Sci. Phila., 1883, p. 32; Stejneger, N. A. Fauna, No. 7, 1893, p. 204; Boulenger, Cat. Snakes Brit. Mus., II, 1894, p. 209.

Hypsiglena chlorophæa, Cope, Proc. Ac. Nat. Sci. Phila., 1860, p. 247 (type locality **Fort Buchanan, Ariz.**); Stejneger, N. A. Fauna, No. 7, 1893, p. 205.

Hypsiglena texana, Stejneger, N. A. Fauna, No. 7, 1894, p. 205 (type locality "between **Laredo and Camargo, Tex.**").

Description.—Head flat-topped or slightly rounded, and snout projecting. Temporal regions usually swollen. Rostral plate large, prominent, recurved on top of snout, and bounded behind by internasal, anterior nasal, and first labial plates. Plates on top of head, a pair of internasals, a pair of prefrontals, a frontal, supraocular of each side, and a pair of rather short, rounded parietals.

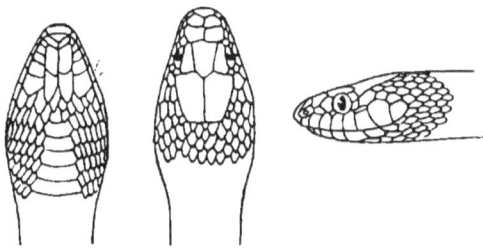

Anterior and posterior nasals usually distinct, but sometimes united above nostril. Loreal small, but often elongate. Two (or one or three) pre- and two postoculars. Temporals normally one followed by two, rarely 2–2. Eight (rarely seven) superior and ten (or nine) inferior labials, sixth or seventh superior and fifth or sixth inferior largest, fourth and fifth or fifth and sixth superior reaching eye, first pair of inferior meeting on median line. Geneials in two pair, posterior equal to or a little larger or smaller than anterior. Scales on body smooth, thin, in twenty-one rows. Anal plate divided. Gastrosteges varying in number from one hundred and sixty-seven to one hundred and eighty-seven. Urosteges in two series of from forty to fifty-five.

The ground color above is yellowish white, so thickly sprinkled with minute brown or black dots as to present an ashy or olivaceous appearance. Along the middle of the back is a single or double series of more or less alternate and confluent blotches of brown or black. On the sides are two or three or four alternating series of small brown or black spots. There are two or three

elongate dark blotches on the nape,
each lateral one being produced
forward as a narrow band across the
side of the face. These nuchal
blotches often unite to form a dark
transverse band or collar. The top
of the head, the labials, and the
geneials are spotted with brown. The gastrosteges are
yellowish or white, immaculate. The urosteges are
sometimes speckled with gray or brown.

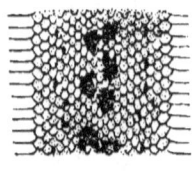

Length to anus.....................137   282   284   318   325   447
Length of tail..................... 21   60   56   47   60   76

Distribution.—This little snake has been taken in
California at San Diego and in the Cuyamaca Mts., San
Diego County, at San Jacinto and Strawberry Valley
(5,000 feet), Riverside County, at Hesperia, San Bernar-
dino County, and in Shepherd Cañon, Argus Range,
Inyo County.

Habits.—Unknown. Probably nocturnal.

Genus 32. SALVADORA.

Salvadora B. & G. Cat. N. A. Rept., I, Serp., 1853, p. 104 (type
grahamiæ); *Phimothyra*, COPE, Proc. Ac. Nat. Sci. Phila.,
1860, p. 566 (type grahamiæ).

The body is very long and slender, with long whip-
like tail. The head is distinct from the neck, large,
long, flat-topped, with truncate snout. Its plates are
normal, except the rostral which is very large and has
free lateral edges. The nasal plates are distinct. Two
preoculars, two postoculars, and a loreal are present.
Temporals are 1-2, 2-2, 2-3, or 3-3. The scales are
smooth, in seventeen rows, with two apical pits. The
anal plate is divided. Urosteges are in two series. The
eye is large, with round pupil.

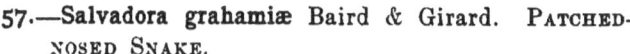

57.—**Salvadora grahamiæ** Baird & Girard. PATCHED-
NOSED SNAKE.

Salvadora grahamiæ, B. & G., Cat. N. A. Rept., Pt. I, Serp., 1853,
   p. 104 (type locality "**Sonora, Mex.**"); BAIRD, U. S. Mex.
   Bound. Surv., II, 1859, p. 21, pl. V, fig. 2; BOCOURT, Miss. Sci.
   au Mex. Rept., 11e Livr., 1888, p. 661, pl. XLIII, figs. 2-2e ;
   COPE, Proc .U. S. Nat. Mus. XIV, 1891 (1892), p. 619; VAN
   DENBURGH, Proc. Cal. Acad. Sci. (2), V, Pt. 1, 1895, p. 146.
Phimothyra grahamiæ, COPE, Proc. Ac. Nat. Sci. Phila., 1861, p.
   300; COPE, l. c., 1883, p. 14.
Phimothyra hexalepis, COPE, Proc. Ac. Nat. Sci. Phila., 1866, p. 304
   (type locality **Fort Whipple, Ariz.**).
Zamenis grahami, BOULENGER, Cat. Snakes Brit. Mus., I, 1893, p.
   393 (part).
Salvadora grahamiæ hexalepis, STEJNEGER, N. A. Fauna, No. 7, p.
   205.

Description.—Head flat-topped or slightly rounded,
and snout projecting and very blunt. Temporal regions
not swollen. Rostral plate very large, prominent, recurved on top of snout, with free lateral edges, and
bounded behind by internasal, anterior nasal, and first
labial plates. Plates on top of head, a pair of internasals, a pair of prefrontals, a frontal, supraocular of
each side, and a pair of rather short, rounded parietals.
Anterior and poste-
rior nasals distinct.
Loreal small, but
sometimes divided.
Two or one pre- and
two or three postocu-
lars. Temporals normally two followed by three,
rarely 1-2, 2-2, or 3-3. Eight or nine superior and
about ten inferior labials, sixth or seventh or seventh
and eighth superior and fifth or sixth inferior largest,
fourth and fifth or fifth and sixth superior, or small detached portions of them, reaching eye, first pair of
inferior meeting on median line. Two pair of geneials,

posterior equal to or smaller than anterior. Scales on
body smooth, thin, in seventeen rows. Anal plate
divided. Gastrosteges varying in number from one
hundred and seventy-four to two hundred and six.
Urosteges in two series of from seventy-four to one
hundred and eight.

The upper surface of the head is grayish or yellowish
brown, without dark or light markings. The ground
color of the body is drab, yellow, light brown, or plumbeous, with one or two narrow black,
slate, brown, or olive longitudinal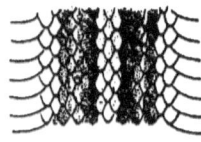
bands. These bands fade out on the
tail, and may extend on to the sides
of the head. Sometimes they widen
and merge; in other specimens the
upper band is partially broken up into spots; while in
one the bands are "represented by blackish shades at
the bases of the scales." The upper lip and all the
lower surfaces are yellow or yellowish white, the tail
and posterior part of the body sometimes with an orange
or reddish cast.

| Length to anus | 235 | 258 | 340 | 652 | 695 | 850 |
| Length of tail | 63 | 60 | 98 | 215 | 189 | 290 |

Distribution.—In California, the Patched-nosed Snake
probably ranges over most of the desert area. It has
been found at Valle de la Viejas, San Diego County,
San Jacinto, Riverside County, at Lytle Creek and near
San Bernardino, San Bernardino County, and in Inyo
County at the Amargosa Borax Works and in the
Argus Range, California. It has been collected in
Nevada at St. Thomas and the Virgin River near
Bunkerville.

Genus 33. BASCANION.

Bascanion, B. & G., Cat. N. A. Rept., Pt. I, Serp., 1853, p. 93 (type constrictor); *Masticophis*, B. & G., l. c., p. 98 (type ornatus).

The body is very long and slender, with long whip-like tail. The head is distinct from the neck, large, long, with flattened top and rounded snout. Its plates are normal. The nasal plates are not united. There are two (rarely one) preoculars and two postoculars. Temporals are normally 2–2. A loreal is present. The scales are smooth, in fifteen, seventeen, or nineteen rows, usually with two (0–3) apical pits. The anal plate is divided.* Urosteges are in two series. The eye is very large, with round pupil.

Four species are known to be Californian. Young of the first two are blotched, of the others, striped.

SYNOPSIS OF SPECIES.

a.—Scales in seventeen rows.
  b.—No distinct longitudinal light lines.
    c.—Urosteges fewer than 100; gastrosteges fewer than 185.
                B. constrictor vetustum.—p. 183.
    c².—Urosteges more than 100; gastrosteges more than 190.
                B. flagellum frenatum.—p. 186.
  b².—A light line along third and fourth rows of scales.
                B. laterale.—p. 188.
a².—Scales in fifteen rows...........**B. tæniatum.**—p. 190.

58.—Bascanion constrictor vetustum Baird & Girard.

WESTERN YELLOW-BELLIED RACER.

> *Bascanion vetustus*, B. & G., Cat. N. A. Rept., Pt. I, Serp., 1853, p. 97 (type locality **San José, Calif.**); GIRARD, U. S. Explor. Exped., Herp., 1858, p. 127, pl. VIII, figs. 12–19.
> *Bascanium constrictor flaviventris*, YARROW & HENSHAW, Surv. W. 100th Mer., App. NN., 1878, p. 213; YARROW, Bull. U. S. Nat. Mus., No. 24, 1882, p. 110 (part).

---

* This plate has been found undivided in *B. flagellum* and its westernmost subspecies *B. f. frenatum.*

Bascanium constrictor vetustum, TOWNSEND, Proc. U. S. Nat. Mus.,
  X, 1887, p. 240.
Bascanium constrictor, COPE, Proc. U. S. Nat. Mus. XIV, 1891 (1892),
  p. 624 (part).
Zamenis constrictor, BOULENGER, Cat. Snakes Brit. Mus., I, 1893,
  p. 387 (part).

Description.—Head rather long, with flattened top and
rounded snout. Rostral plate large, about as high as
wide, hollowed below, and bounded behind by internasal, anterior nasal, and first labial plates. Plates on
top of head, a pair of internasals, a pair of prefrontals,

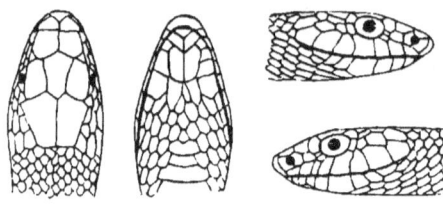

supraocular and part of upper preocular of each side,
long frontal, and a pair of large parietals. Anterior
and posterior nasals distinct. Loreal well developed.
Preoculars normally two, but sometimes united. Postoculars two, upper a little larger than lower. Temporals
normally two followed by two, but may be 2–3, 1–2, or
1–1. Seven or eight superior and eight or nine inferior
labials, next to last of upper and fifth (or fourth) of
lower largest, third and fourth or fourth and fifth
superior reaching eye, first pair of inferior meeting on
median line. Geneials in two pair, equal or either pair
a little larger. Scales on body smooth, in seventeen
rows. Anal plate divided. Gastrosteges varying in
number from one hundred and sixty-three to one hundred and seventy-nine. Urosteges in two series of
from seventy-nine to ninety-eight.

The color above in adults is green, olive, or yellowish or reddish brown, changing to green (or blue) on the lower rows of scales and the tips of the gastrosteges. There are no dark or light markings but the skin between the scales is often 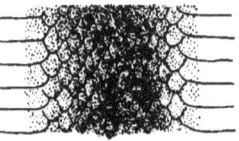 black. The head and tail are unicolor with the body. The lower surfaces are yellow or, rarely, white, unspotted.

Young are colored like adults on the tail and posterior part of the back, but anteriorly are spotted, blotched, or cross-barred with brown of a shade darker than the ground-color. These dark markings spread and blend until the adult coloration is 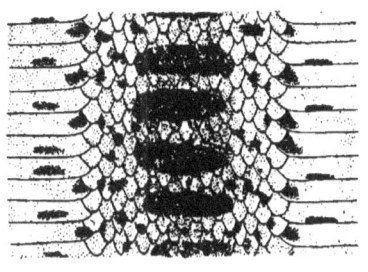 assumed. Dark spots are present also on the tips of the gastrosteges and sides of the head.

Length to anus......................203	314	526	626	636	647
Length of tail...................... 64	103	196	217	192	209

Distribution.—The Western Yellow-bellied or "Blue" Racer ranges over the whole length of California, but, I believe, has never been taken in the desert regions of the southeast. It has been collected in San Diego (Agua Caliente 3,400 feet), San Bernardino, Kern (Fort Tejon, Kernville), Monterey (Monterey), Santa Cruz (Glenwood), Santa Clara (Los Gatos, San José, Palo Alto), San Mateo (Pescadero), San Francisco, Alameda (Berkeley), Contra Costa (Crockett), Marin

(Mill Valley, Camp Taylor), Sonoma (Healdsburg),
Mariposa (Yosemite Valley), El Dorado (5,000 feet),
Placer (Red Point), Lassen (Honey Lake), and Shasta
(McCloud River) Counties, California. It crosses Ore-
gon (Klamath Lake, Summer Lake, Warner's Valley,
Willamette Valley, Dalles) to Washington (Ft. Steila-
coom, Puget Sound) and Idaho (Atlanta, mouth of Bru-
neau River, Big Butte).

Habits.—Like other members of its genus, the Western Yellow-bellied Racer is a skillful climber and often
runs through the tops of the bushes at almost as great
a speed as when upon the ground. It is frequently
found, however, in open country or in fields of growing
grain.

59.—**Bascanion flagellum frenatum** Stejneger. WESTERN
WHIP SNAKE.

> *Bascanium flagelliforme testaceum*, YARROW, Bull., U. S. Nat. Mus.,
> 24, 1882, p. 112 (part).
> *Bascanium testaceum*, COPE, Proc. Ac. Nat. Sci. Phila., 1883, pp.
> 29, 32.
> *Bascanium flagelliforme*, COPE, Proc. U. S. Nat. Mus., XIV, 1891
> (1892), p. 625 (part).
> *Bascanion flagellum frenatum*, STEJNEGER, N. A. Fauna, No. 7,
> 1893, p. 208 (type locality **Mountain Spring, Colorado
> Desert, San Diego County, Calif.**)

Description.—Head rather long, with flattened top, and
narrow, rounded snout. Rostral plate large, high, hollowed below, and bounded behind by internasal, anterior nasal, and first labial plates. Plates on top of head,
a pair of internasals, a pair of prefrontals, supraocular
and part of upper preocular of each side, long frontal,
and pair of large parietals. Anterior and posterior
nasals distinct. Loreal well developed. Preoculars
normally two, but sometimes united. Postoculars two,
upper a little larger. Temporals normally two followed

by two, but rarely 2–1, 1–2, or 1–1.   Eight or nine
superior and ten or eleven inferior labials, seventh or

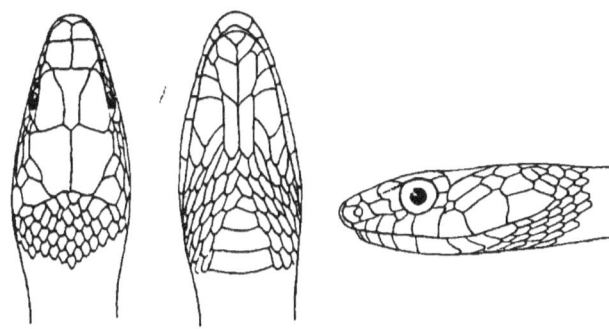

eighth upper and fifth lower largest, fourth and fifth or
fifth and sixth superior reaching eye, first pair of inferior meeting on median line.   Geneials in two pair,
posterior pair a little larger.   Scales on body smooth,
in seventeen rows.   Anal plate almost always divided.
Gastrosteges varying in number from one hundred and
ninety-seven to two hundred and six.   Urosteges in two
series of from one hundred and four to one hundred
and twenty-three.   Third, fourth, and fifth urosteges of
one specimen not divided.

The general color is whitish, grayish, ochraceous,
brownish, or straw yellow, usually lightest at the edges
of the scales, often spotted with brown or black at their
tips or bases.   Across the
nape are several (3–7)
brownish or blackish
bands, often more or
less blended.   Faint in-
dications of longitudinal
lines may sometimes be
seen along the sides.   The lower surfaces are pale

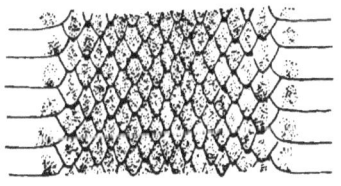

yellow or white, more or less spotted anteriorly with black, gray, brown, or yellow. These spots usually form one row along each side of the anterior gastrosteges.

Young are more or less distinctly cross-barred above with gray, brown, or black, and show a light line or blotch along the side of the face. The latter mark is often retained by adults.

Length to anus.................................303   541   723
Length of tail................................. 96   185   242

Distribution.—The Western Whip Snake or "Red Racer" has its true home in the deserts of San Diego, San Bernardino, Kern, and Inyo Counties, but lives also in the southwestern part of the State, and along the eastern side of the San Joaquin Valley. It has been taken in San Diego (Fort Yuma, Mountain Spring Colorado Desert, San Diego, Agua Caliente), Riverside (Palm Springs, San Jacinto, Riverside), Los Angeles (Pasadena, Drum Barracks), San Bernardino (Ontario, Needles), Inyo (Death Valley, Panamint Valley, Deep Spring Valley, Keeler, Owen's Valley), Fresno (Fresno), Mariposa (Yosemite Valley) Counties, California, and crosses southern Nevada (Vegas Valley, Overton).

Habits.—The "Red Racer," like its relatives, is remarkable for the quickness of its movements. It climbs trees and bushes with great agility.

60.—Bascanion laterale (Hallowell). CALIFORNIA RACER.

Leptophis lateralis, HALLOW., Proc. Ac. Nat. Sci. Phila., VI, 1853, p. 237 (type locality California); HALLOW., Pac. R. R. Surv., X, Rept., 1859, p. 13, pl. IV, fig. 3a–3c.
Bascanium tæniatum laterale, YARROW, Bull. U. S. Nat. Mus., 24, 1882, p. 113 (part).
Bascanium laterale laterale, COPE, Proc. U. S. Nat. Mus., XIV, 1891, (1892), p. 628.
Bascanion laterale, STEJNEGER, N. A. Fauna, No. 7, 1893, p. 209.

Description.—Head long, with flattened top and narrow, rounded snout. Rostral plate large, about as high as broad, hollow below, and bounded behind by internasal, anterior nasal, and first labial plates. Plates on top of head, a pair of internasals, a larger pair of prefrontals, supraocular and part of upper preocular of each side, long and posteriorly narrow frontal, and a pair of very large parietals. Anterior and posterior nasals distinct. Loreal rather large, rarely with a small plate below it. Preoculars two, upper much larger than lower. Postoculars two (or three), nearly equal. Temporals two followed by two, rarely 1–2 or 1–1. Eight (or seven) superior and ten (or nine) inferior labials, seventh (or sixth) of upper and fifth of lower series largest, fourth and fifth (or third and fourth) of upper reaching eye, first pair of lower meeting on median line. Geneials in two pair, posterior larger than anterior. Scales on body smooth, in seventeen rows. Anal plate divided. Gastrosteges varying in number from one hundred and ninety to one hundred and ninety-seven. Urosteges in two series of from one hundred and twenty-two to one hundred and thirty-three. Tail very long and slender.

The color above, including the tips of the gastrosteges and urosteges, is dark brown, palest on the tail. A single light yellow or white line extends along each side, on the third and fourth rows of scales, to, or a little beyond, the base of the tail. This line is often bordered with black and its scales are sometimes tipped with orange-rufous. The  sides of the head are spotted with yellow. All the lower surfaces and the upper labials are yellow, spotted on the head and anterior gastrosteges with gray, slate, or brown. The anterior gastrosteges are sometimes washed with orange-rufous.

Length to anus......632	725	742	787	804	968
Length of tail......275	311	329	338	396	

Distribution.—The Californian Racer appears to be confined to those parts of California which lie to the west of the Sierra Nevada and Mojave and Colorado Deserts. The northern limit of its range is not known. It has been taken in San Diego (Santa Ysabel, Agua Caliente, Oak Grove), Riverside (San Jacinto, Strawberry Valley, Riverside), Los Angeles (Los Angeles), Santa Barbara (Santa Barbara), Kern (Fort Tejon, Walker Pass), Tulare (Three Rivers), Fresno (Fresno), Monterey (Carmel Valley), Santa Clara (Los Gatos), and Lake (Mt. Saint Helena) Counties.

Habits.—Nothing is known of the habits of this snake, except that, like other members of the genus, it is very active and a skillful climber.

61.—Bascanion tæniatum (Hallowell). STRIPED RACER.

Leptophis tæniata, HALLOW., Proc. Ac. Nat. Sci. Phila., 1852, p. 181 (type locality New Mexico); HALLOW., Sitgreave's Zuni and Colorado Riv., 1853, p. 133, pl. XIII (XII).

Masticophis tæniatus, BAIRD & GIRARD, Cat. N. A. Rept., Pt. I, Serp., 1853, p. 103; B. & G., Pac. R. R. Surv., X, Rept., 1859, pl. XXXII, fig. 76.

Bascanium tæniatum tæniatum, YARROW, Bull. U. S. Nat. Mus., 24, 1882, p. 112.

Bascanium tæniatum, COPE, Proc. U. S. Nat. Mus., XIV, 1891 (1892), p. 629.

Bascanium tæniatum, STEJNEGER, N. A. Fauna, No. 7, 1893, p. 210.

Zamenis tæniatus, BOULENGER, Cat. Liz. Brit. Mus., I, 1893, p. 390 (part).

Description. —Head long, with flattened top and rounded snout. Rostral plate large, about as high as

broad, hollowed below, and bounded behind by internasal, anterior nasal, and first labial plates.   Plates on
top of head, a pair of interna-
sals, a pair of prefontals, supraocular and part of upper preocular of each side, long frontal, and a pair of large parietals.
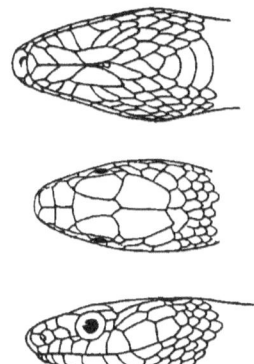
Anterior and posterior nasals
distinct.   Loreal longer than
high.   Preoculars two, upper
much larger than lower.   Postoculars two, nearly equal.
Temporals two followed by
two.   Eight superior and nine
or ten inferior labials, sev-
enth of upper and fifth of lower largest, fourth and
fifth upper reaching eye, first pair of lower meeting on median line.   Geneials in two pair, anterior
smaller than posterior.   Scales on body smooth, in fifteen rows.   Anal plate divided.   Gastrosteges varying
in number from two hundred and four to two hundred
and nine.   Urosteges in two series of from one hundred
and thirty-six to one hundred and fifty-seven.

   The color above is grayish or yellowish brown, palest
on the tail, the scales of the sides more or less yellow
and with narrow dark lines along
the middle of each row.   These dark

lines are sometimes present on the
dorsal scales as well as those of the
sides and a wider line often runs
along the tips of the gastrosteges.   The yellow of
the lateral scales is variable in intensity, and sometimes—especially in young—forms a distinct line
along the third and fourth rows of scales.   All of these
lines fade out on the tail.   The head is spotted with

yellow. The lower surfaces are yellow or yellowish
white, marked with slate or black anteriorly and along
the tips of the gastrosteges, and often more or less tinted
posteriorly with delicate rose pink.

Length to anus...803 – 910
Length of tail...350 – 366 +

Distribution.—The Californian range of this racer
seems to be restricted to the dryer eastern portions of
the State, leaving the more western parts to *B. laterale.*
" It is much more widely distributed [than the Californian racer], as specimens have been taken in Idaho,
Utah, Nevada, Arizona, New Mexico, and Mexico, but
it does not seem to reach the coast, nor does it appear
in the Valley of California, except at two points. These
are Walker Basin and Shasta County, northern California, where [Shasta Co.] it probably enters by way of Pitt
River Valley."* Localities at which it has been taken
in California are Shasta County (Baird, Canoe Creek),
Inyo County (Argus Range, Coso Valley and Mountains,
Panamint Mountains), and Kern County (Walker
Basin); in Nevada, Carson City, and Antelope Springs;
in Oregon, Snake River; and in Idaho, between Bliss
and the Snake River.

Genus 34. ARIZONA.

Arizona, KENN., U. S. Mex. Bound. Surv., II, 1859, Rept., p. 18
(type **elegans**).

The body is long and slender, with tail of moderate
length. The neck is constricted somewhat, so that the
head is distinct. The snout is long, rounded, and but
little lower than the flat top of the head. The cephalic
plates are normal. The nasals rarely unite above
the nostril. One (or two) preocular, two (or one)

* Stejneger, N. A. Fauna, No. 7, 1893, p. 210.

postoculars, and a loreal are present. Temporals are 2-3 or 2-4. The scales are smooth, in twenty-seven to thirty-one rows. The anal plate is single. Urosteges are in two series. The eye is moderately large, with round pupil.

62.—Arizona elegans Kennicott. FADED SNAKE.

Arizona elegans, KENN., U. S. Mex. Bound. Surv., II, 1859, Rept., p. 18, pl. XIII (type locality **Rio Grande**); BOCOURT, Miss. Sci. au Mex., Rept., 11e Livr., 1888, p. 676, pl. XLVI, figs. 3-3b.

Pityophis elegans, COPE, Bull. U. S. Nat. Mus., No. 1, 1875, p. 39.

Rhinechis elegans, COPE, Proc. Am. Philos. Soc., XXIII, 1886, p. 284; COPE, Proc. U. S. Nat. Mus., XIV, 1891 (1892), p. 638.

Coluber arizonæ, BOULENGER, Cat. Snakes Brit. Mus., II, 1894, p. 66.

Description.—Head flat-topped or slightly rounded, with snout projecting and rather narrow. Temporal regions not swollen. Rostral plate very large, prominent, recurved between internasals on top of snout, and bounded behind by internasal, anterior nasal, and first labial plates. Plates on top of head, a pair of internasals, a pair of prefrontals, a frontal, supraocular of

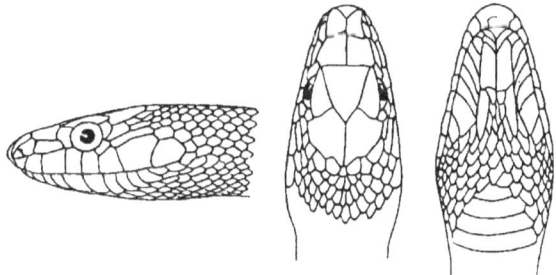

each side, and a pair of large parietals. Anterior and posterior nasals usually distinct, but sometimes united above nostril. Loreal elongate. One or two preoculars and two or one postoculars. Temporals normally two followed by three, or 2-4, lower scale of first row often

elongate.   Eight or seven superior and thirteen to
fifteen inferior labials, sixth or seventh superior and
seventh or eighth inferior largest, fourth and fifth
superior reaching eye, first pair of inferior meeting on
median line.   Two pair of geneials, posterior narrower
and shorter or but little longer than anterior.   Scales
on body smooth, thin, in twenty-seven to thirty-one
rows.   Anal plate not divided.   Gastrosteges varying
in number from two hundred and ten to two hundred
and twenty-seven.   Urosteges in two series of from
forty-five to fifty-nine.

The ground color is pale brown or yellowish gray,
lighter near the middle of
the back, along which is a
series of dark-edged brown
or gray blotches.   The
sides are marbled with
similar but smaller
blotches in more or less
alternating rows.   The
tail is similarly colored.

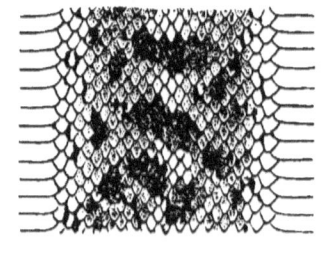

In young, a dark streak runs from the eye to the corner
of the mouth, and another between the eyes, crossing
the posterior edges of the prefrontal plates as in *Pituophis*.   These streaks are faded or absent in adults.   The
lower surfaces are pale yellowish white, without markings.

Length to anus.........................231   282   730   860   870
Length of tail......................... 33   33   122   160   125

Distribution.—This smooth-scaled relative of the true
gopher snakes has been taken in California between
Carlsbad and Oceanside and at Warner's Ranch, San
Diego County, at San Jacinto, Riverside County, and
near Ontario, San Bernardino County.   It ranges east
to Texas.

Habits.—Unknown. A captive individual ate a Brown-shouldered Lizard *(Uta stansburiana)*.

Genus 35. PITUOPHIS.

Pituophis, HOLBROOK, N. A. Herpet. (2), IV, 1842, p. 7 (type mel-anoleucus); *Churchillia*, B. & G., Stansbury's Exped. Gt. Salt Lake, 1852, p. 350 (type bellona).

The body is long but rather stout, with tail of moderate length. The neck usually is slightly constricted, so that the head appears little distinct from it. The snout is long, narrowly rounded, and projecting beyond the lower jaw. The head-plates show many variations, but when typical are normal except that there are four prefrontals. The nasals are usually distinct. One or two preoculars, two to four postoculars, and a loreal are present. Temporals are many and very variable. The scales are in twenty-seven to thirty-five rows, the dorsals keeled, some of the lateral rows smooth. The anal plate is single. Urosteges are in two series. The eye is large, with round pupil.

63.—Pituophis catenifer (Blainville). WESTERN GOPHER SNAKE.

Coluber catenifer, BLAIN., Nouv. Ann. Mus. Hist. Nat., IV, 1835, p. 290, pl. 26, figs. 2-2b (type locality California); BOULENGER, Cat. Snakes Brit. Mus., II, 1894, p. 67.

Pituophis catenifer, B. & G., Cat. N. A. Rept., Pt. I, Serp., 1853, p. 69; GIRARD, U. S. Explor. Exped., Herp., 1858, p. 135, pl. VIII, figs. 1-7; GÜNTHER, Cat. Colub. Snakes Brit. Mus., 1858, p. 87; STEJNEGER, N. A. Fauna, No. 7, 1893, p. 206.

Pituophis Wilkesii, B. & G., Cat. N. A. Rept., Pt. I, Serp., 1853, p. 71 (type locality "Puget Sound, Oregon"); GIRARD, U. S. Explor. Exped., Herp., 1858, p. 137, pl. IX, figs. 1-7.

Pituophis annectens, B. & G., Cat. N. A. Rept., p. 72 (type locality San Diego, Cal.).

Pityophis Heermannii, HALLOW., Proc. Ac. Nat. Sci. Phila., VI, 1853, p. 236 (type locality Cosumnes River, Cal.).

Pityophis catenifer, COPE, Proc. U. S. Nat. Mus., XIV, 1891, p. 641.

Description.—Head flat-topped or rounded, with snout projecting and rather narrow. Temporal regions not swollen. Rostral plate very large, prominent, not very narrow, but often recurved between internasals on top of snout, and bounded behind by inter- 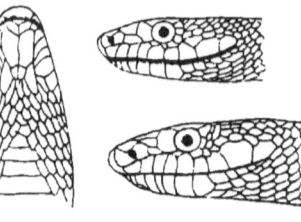 nasal, anterior nasal, and first labial plates. Plates on top of head, a pair of internasals, four, two, one, three, five, six, or eight prefrontals, a frontal, supraocular of each side, and a pair of parietals. Anterior and posterior nasals usually distinct, sometimes united. Loreal usually elongate. One or two pre- 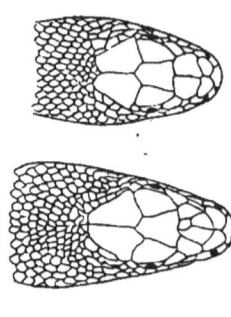 oculars and two to four postoculars. Tips of some labials often cut off, forming suboculars or loreals. Temporals four followed by five, very variable. Seven to nine superior and eleven to thirteen inferior labials, next to last superior and sixth or seventh inferior largest in its series, fourth or fifth 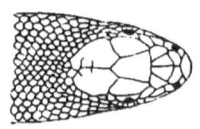 superior usually reaching eye, first pair of inferior meeting on median line. Two pair of genials, anterior much larger than posterior. Scales on body in twenty-seven to thirty-five rows, keeled except in a varying number (about 3 to 10) of rows on each side. Anal plate not divided. Gastrosteges varying in number from two hundred and four to two hundred and forty-five. Urosteges in two series of from fifty-six to seventy-two.

The ground color is pale brown or yellowish gray, sometimes obscured by the spreading of the blotches or the presence of black marks along the keels of its scales. Along the middle of the back is a series of from thirty-six to seventy-nine dark brown or black 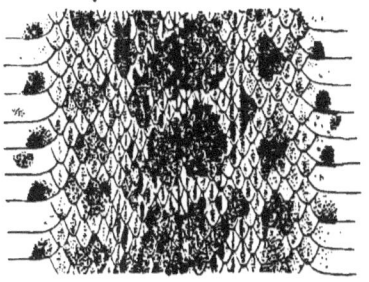 blotches. There are several series of smaller alternating dark blotches or spots on the sides. These spots sometimes tend to unite to form longitudinal lines. Across the top of the head, between the preocular plates, is a line of black or brown. A similar line runs down from the center of the eye and another back and down from the upper postocular plate. The lower surfaces are yellowish white, usually maculated with black or brown.

Length to anus.....................315   660   820   860   1200   1260
Length of tail..................... 52   135   165   170   270   210

Distribution.—The Western Gopher Snake occupies the Pacific Coast of the United States, but is replaced in the Great Basin by its subspecies *P. catenifer deserticola.* It ranges north and south at least from Puget Sound to San Diego. The eastern limit of its territory in the north (in Idaho) is not known, but in California is marked by the Sierra Nevada and the western edge of the Mojave and Colorado Deserts. In California, it has been taken in San Diego (San Diego), Riverside (San Jacinto), San Bernardino (Ontario), Los Angeles (Pasadena), Santa Barbara (Santa Barbara), Kern (Fort Tejon), Fresno (Pitman Creek, King's River), Montery (Monterey), Santa Cruz (Soquel, Corralitos, Glenwood), Santa

Clara (Los Gatos, San José, Palo Alto), San Francisco, Alameda (Haywards, Oakland), Marin (Tamalpais), Sonoma (Petaluma), Lake (Kelseyville), Mendocino, Shasta (McCloud River), and Siskiyou (Mt. Shasta) Counties.

Habits.—The Gopher or Bull Snake is the largest as well as one of the most abundant of Californian serpents. Individuals more than six feet long are not rarely found. These are usually very gentle and show little resentment even when roughly handled. The younger snakes, however, sometimes strike very fiercely, but of course harmlessly. This snake shares with many others the curious habit of rapidly vibrating the tip of its tail when excited; an action which sometimes, when the tail happens to strike upon dry leaves or grass, produces a sound not unlike the warning whir of the rattlesnake. Its food, so far as is known, consists of small mammals, of which gophers are said to form a large part.

64.—Pituophis catenifer deserticola Stejneger. DESERT GOPHER SNAKE.

Pityophis sayi bellona, COPE, Proc. U. S. Nat. Mus., XIV, 1891, p. 641 (part).

Pituophis catenifer deserticola, STEJNEGER, N. A. Fauna, No. 7, 1893, p. 206 (type locality **Great Basin and southwestern deserts**).

Description.—I have seen no specimens of this "richly-colored form from the Great Basin and southwestern deserts, which agrees with true *P. catenifer* in having a broad and low rostral. * * *' As a general rule this form has a more pronounced carination of the scales and a less number of smooth scales on the sides, but this character cannot be relied upon at all, and whether a specimen shall be referred to either typical *P. catenifer* or to this desert form must be decided upon the totality of the characters, as a reliance upon the carination leads to very erroneous results."

Distribution.—The Desert Gopher Snake occupies the Mojave and Colorado Deserts and those parts of the Great Basin which lie within California, Nevada, and perhaps southern Idaho and eastern Oregon.

Habits.—Unknown, but similar, doubtless, to those of *P. catenifer.* :

Genus 36. THAMNOPHIS.·

Thamnophis, FITZINGER, Syst. Rept., 1843, p. 26 (type **saurita**); *Eutainia,* B. & G., Cat. N. A. Rept., Pt. I, Serp., 1853, p. 24 (type **saurita**).

The body is more or less elongate, usually rather slender, with moderately long, tapering tail, and head distinct from neck. The cephalic plates are normal. The nasals never unite. One or two (rarely three) pre-oculars, two to four postoculars, and a loreal are present. Temporals are normally 1-2, but may be 1-1, 1-3, or 2-3. A loreal is present. The scales are keeled, in seventeen to twenty-three rows. The anal plate is undivided. Urosteges are in two series. The eye is moderate or small, with round pupil.

Although the several species and subspecies may often be distinguished at a glance by one familiar with their several characters, the amount of individual variation is so great that it is quite impossible to make a key which will properly refer all specimens. The following synopsis will, I believe, usually serve its purpose, but should not be trusted too implicitly.

SYNOPSIS OF SPECIES.*

a.—Upper labials seven; posterior geneials much longer than anterior.

---

* If my estimate of the status of the various western members of this genus differs rather radically from that of recent authors (the latest American review of this genus admits seventeen species and subspecies from the territory under discussion), I may, perhaps, be pardoned on the ground that I have carefully examined more than three hundred fresh alcoholic specimens besides rather hastily inspecting the material in the National Museum.

b.—Eye large; one preocular; scales in nineteen rows; dorsal line never red.

    c.—Lines wide; some red on body.....**T. parietalis.**—p. 200. ✓ ———

    c².—Lines very narrow; no red on body; belly dusky.

        T. p. pickeringii.—p. 204.

b².—Eye small; two preoculars or scales in seventeen rows, or both; dorsal line sometimes red..................**T. leptocephala.**—p. 205.

a².—Upper labials eight.

    d.—Scales in nineteen rows (if in 21 rows, some red on body or very distinct lines with few or no spots).

        e.—Eye smaller; posterior geneials rarely much longer than anterior; dorsal line yellow or red; head never red; belly sometimes red.

            T. elegans.—p. 207.

        e².—Eye larger; posterior geneials much longer than anterior; dorsal line blue or yellow; head sometimes red; belly never red.

            T. parietalis.—p. 200.

    d².—Scales in twenty-one (or twenty-three) rows; no red on body; many spots or no distinct dorsal line.

        f.—Eye smaller; posterior geneials shorter or but little longer than anterior; belly marbled with slate; dorsal line present.

            g.—One preocular; scales in twenty-one rows.

                T. vagrans.—p. 210.

            g².—Two preoculars; scales in twenty-one or twenty-three rows.

                T. v. biscutata.—p. 212.

        f².—Eye larger; posterior geneials much longer than anterior; belly not marbled with dusky; dorsal line often absent.

            T. hammondii.—p. 212. ✓

65.—Thamnophis parietalis (Say). PACIFIC GARTER SNAKE.

Coluber parietalis, SAY, Long's Exped. Rocky Mts., I, 1823, p. 186 (type locality **Camp Missouri near Council Bluff**).

Coluber infernalis, BLAINV., Nouv. Ann. du Mus., IV, 1835, p. 291, pl. XXVI, figs. 3-3ᵃ. (type locality **California**).

Eutainia infernalis, B. & G., Cat. N. A. Rept., Pt. 1, Serp., 1853, p. 26; GIRARD, U. S. Explor. Exped., Herp., 1858, p. 148, pl. XIV, figs. 11-16; BOCOURT, Bull., Soc. Zool. Fr., 1892, p. 40 (part).

Eutainia parietalis, B. & G., Cat. N. A. Rept., Pt. 1, Serp., 1853, p. 28.

?*Eutaenia ornata*, B. & G., U. S. Mex. Bound. Surv., III, Rept., 1859, p. 16, pl. IX (type locality between **San Antonio** and **El Paso, Tex.**).

Eutænia sirtalis tetratænia, COPE, U. S. Explor. Surv. W. 100th
Mer., V, 1875, p. 546 (no locality); COPE, Proc. U. S. Nat.
Mus., XIV, 1891, p. 664 (**Puget Sound, Wash.,** and **Pitt
River, Cal.**).

Eutænia pickeringii, COPE, Proc. Ac. Nat. Sci. Phila., 1883, p. 21.

Tropidonotus sirtalis var. *parietalis,* GARMAN, Mem. Mus. Compar.
Zool., VIII, 3, 1883, pp. 25, 139.

Eutænia sirtalis parietalis, COPE, Proc. U. S. Nat. Mus., XIV, 1891,
p. 664.

Eutænia sirtalis trilineata, COPE, Proc. U. S. Nat. Mus., XIV, 1891,
p. 665 (part), (**Fort Benton, Mont.**).

Tropidonotus ordinatus var. *infernalis,* BOULENGER, Cat. Snakes Brit.
Mus., I, 1893, p. 207 (part).

Thamnophis parietalis, STEJNEGER, N. A. Fauna, No. 7, 1893, p. 214.

Description.—Head distinct from neck, flat-topped,
with narrow, rounded snout, and temporal regions sometimes swollen. Rostral plate large, bounded behind by
internasal, anterior nasal, and first labial plates. Plates
on top of head, a pair of internasals, a pair of prefrontals, a frontal, supraocular of each side, and a pair
of parietals. Anterior and posterior nasals distinct.
One loreal. One preocular and from two to four postoculars. Temporals normally one followed by two, sometimes 1-1, 1-3, or 2-3. Seven, or rarely eight, superior
and nine to eleven inferior labials, fifth or sixth superior and fifth, sixth, or seventh inferior largest, third
and fourth or fourth and fifth superior reaching eye,
first pair of inferior meeting on median line. Two pair
of genials, posterior longer than anterior. Scales on
body in nineteen, or very rarely twenty-one, rows, all
keeled except sometimes the first row of each side.
Anal plate not divided. Gastrosteges varying in number from one hundred and fifty-one to one hundred and
seventy. Urosteges in two series of from seventy-four
to ninety-four, a few of the anterior rarely undivided.
Eye large.

There are three light lines, one on the middle of the

back and one along the second and third rows of scales
of each side, but the lateral lines not infrequently blend
with the color of the belly. The dorsal line usually is
bluish but may be yellow. The belly is bluish or yellowish, or rarely slaty, and may have a black line or
series of spots near the tips of the gastrosteges. The
head may be brown, olive, or coppery red above, bluish
or grayish laterally, yellowish white below. The tail is
colored like the back, but less definitely. In some
specimens the ground color above is solid black, without a trace of red. In others there are traces of red on
the sides, chiefly on the skin between the scales. In
several the red is more extensive and forms small irregular blotches on the sides. In a number these blotches
are larger and extend up from the
lateral line in definite and more or
less rectangular figures, between
which are similarly shaped prolongations downward of the black ground.*
Many show the red blotches spread out and blended
above, so that the downward prolongations of the ground
color have become detached and form a series of black
spots separated, by red,
from the narrow band of
ground color remaining on
each side of the light dorsal line. In others these
black spots have become united and form a black line,
so that on each side of the light dorsal line we have a
line of black, one of red, another of black, and the
light lateral line.† In one specimen the black is almost

entirely replaced by red. These color variations are all
individual, none geographical.

Length to anus	185	475	565	695	715	870
Length of tail	67	160	190	227	200	210 +

Distribution.—The Pacific Garter Snake ranges all
over the territory under consideration, excepting the
desert areas and the western parts of Washington and
Oregon. It is abundant in eastern Washington and
Oregon and in Idaho, and has been reported from several localities in Nevada.

It occurs in all parts of California except the Mojave
and Colorado Deserts. I have examined specimens from
Siskiyou (Mt. Shasta), Placer (Lake Tahoe), Mariposa
(Yosemite Valley), Fresno (Fresno), Humboldt, Mendocino (Pieta), Lake (Kelseyville), Sonoma (Healdsburg, Duncan's Mills), Marin (Tomales Bay), Santa
Clara (Palo Alto), Santa Cruz (Glenwood), San Bernardino (Ontario) and Riverside (Riverside) Counties.

Habits.—Like its relatives, the Pacific Garter Snake is
seldom found far from water. Its food is composed of
fish, batrachians, and the smaller mammals. "Along
the shores of the larger island in Pyramid Lake vast
numbers of *Eutœniœ* are found, comprising this and, in
all probability, several other recognized varieties. During the heated part of the day, the mossy tracts in the
tepid, shallow water of the little inlets were thronged
with them, as they swam in gentle undulations over the
smooth surface or idly basked on the heated rocks along
the shore. In no other locality have we ever seen them
in such numbers. When disturbed, they swam boldly
out into open water or sought the bottom and hid themselves under the rocks. Though not in the true sense
of the word ' water snakes,' the various *Eutœniœ* are all
thus quite aquatic in their habits." *

* Yarrow & Henshaw, Ann. Rep. Surv. W. 100th Mer., Append. NN, 1878, p. 217.

66.—Thamnophis parietalis pickeringii (Baird & Girard).
NORTHWESTERN GARTER SNAKE.

Eutainia Pickeringii, B. & G., Cat. N. A. Rept., Pt. 1, Serp., 1853,
    p. 27 (type locality **Puget Sound**); GIRARD, U. S. Explor.
    Exped., Herp., 1858, p. 150, pl. XIII, figs. 14–20.
Eutainia concinna, B. & G., Cat. N. A. Rept., Pt. 1, Serp., 1853,
    p. 146.
Eutænia sirtalis concinna, COPE, Proc. U. S. Nat. Mus., XIV, 1891,
    p. 664.
Eutænia sirtalis pickeringii, COPE, Proc. U. S. Nat. Mus., XIV, 1891,
    p. 665.
Eutænia sirtalis trilineata, COPE, Proc. U. S. Nat. Mus., XIV, 1891,
    p. 665 (part), ("**Port Townsend, Oregon**").

Description.—This subspecies differs from *T. parietalis*
in color only. Some specimens approach the typical form
in having red blotches on the sides. The normal color-

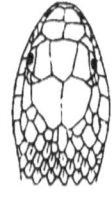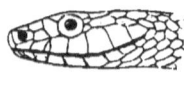

ation is, however, fairly constant. The ground color
above is deep blackish brown or black, with three light
longitudinal lines. These lines are usually very narrow,
but may be as wide as in the typical
T. parietalis, and may be white,
grayish, bluish, greenish, or pale
yellow. The median dorsal line may
fade out posteriorly, and the lateral
lines may be very faint. A narrow
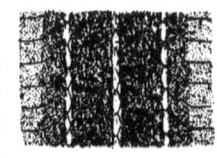
black line sometimes runs along the tips of the gastro-
steges. The top of the head is blackish, its sides light.
The lower surfaces are bluish black, slate, or greenish,
lighter anteriorly; the throat being yellowish white.

This form may be readily distinguished from *T. lepto-cephala*, of the same regions, by its larger eye.

| Length to anus | 165 | 410 | 500 | 540 | 650 | 690 |
| Length of tail | 51 | 125 | 180 | 160 | 190 | 215 |

Distribution.—This northern form of the Pacific Garter Snake inhabits British Columbia and the northwestern corner of the United States in the neighborhood of Puget Sound. It is abundant in King County, Washington, and on Vancouver Island, and has been reported from Oregon.

67.—Thamnophis leptocephala (Baird & Girard). PUGET GARTER SNAKE.

> *? Tropidonotus ordinoides*, B. & G., Proc. Ac. Nat. Sci. Phila., VI, 1852, p. 176 (type locality **Puget Sound**).
> *? Tropidonotus concinnus*, HALLOW., Proc. Ac. Nat. Sci. Phila., VI, 1852, p. 182 (type locality **Oregon Territory**).
> *Eutainia leptocephala*, B. & G., Cat. N. A. Rept., Pt. I, Serp., 1853, p. 29 (type locality **Puget Sound**); GIRARD, U. S. Explor. Exped., Herp., 1858, p. 151, pl. XIII, figs. 7–13.
> *?? Eutainia atrata*, KENN., U. S. Pac. R. R. Surv., XII, Pt. II, 1860, p. 296 (type locality "**California**"?).
> *Eutainia cooperi*, KENN., U. S. Pac. R. R. Surv., XII, Pt. II, 1860, p. 296, pl. XV, fig. 1 (localities **Cathapoot'l and Willopah Valleys**).

Description.—Head distinct from neck, flat-topped, with narrow, rounded snout, and temporal regions sometimes slightly swollen. Rostral large, bounded behind by interna-sal, anterior nasal, and first labial plates. Plates on top of head, a pair of internasals, a 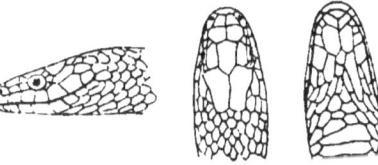 pair of prefrontals, a frontal, supraocular of each side, and a pair of parietals. Anterior and posterior nasals distinct. One loreal. One or two preoculars and about

three postoculars. Temporals normally one followed by
two, sometimes 1–3. Seven, or rarely six or eight, superior and eight to ten inferior labials, fifth or sixth
superior and fourth, fifth, or sixth inferior largest, third
and fourth or fourth and fifth superior reaching eye,
first pair of inferior meeting on median line. Two
pair of geneials, posterior longer than anterior. Scales
on body in seventeen or nineteen rows, all keeled except
sometimes the first row of each side. Anal plate not
divided. Gastrosteges varying in number from one
hundred and thirty-nine to one hundred and fifty-two.
Urosteges in two series of from fifty-two to seventy.
Eye small.

The ground color above is olive or pale blackish
brown, dotted and spotted with black along the edges of
the scales, and with or without three light longitudinal
lines. The lateral lines, when present, are usually
grayish or greenish blue, while the dorsal line—which
often fades out posteriorly—may be white, gray, blue,
yellow, or brick red. A blackish streak usually runs
back from the eye in specimens light enough to show it.
The labials are bluish gray or yellowish. The pineal
spot is often present on the suture between the parietal
plates. The belly may be yellowish, olive, plumbeous,
or slate; the throat is yellowish white; the lower surface
of the tail is sometimes brick red. Many specimens
can be distinguished from *T. p. pickeringii* only by the
smallness of the eye, which, however, is a very good
character.

| Length to anus | 372 | 395 | 420 | 518 | 560 | 585 |
| Length of tail | 105 | 112 | 115 | 133 | 164 | 138 |

Distribution.—This beautiful little snake is very common in British Columbia and Washington in the vicinity of Puget Sound, especially on Vancouver Island and

in King County. It has been noted from western Oregon and from California. The last locality needs confirmation, although this snake may, perhaps, occur in the northwestern corner of our State.

68.—Thamnophis elegans (Baird & Girard). ELEGANT GARTER SNAKE.

Eutainia ordinoides, B. & G., Cat. N. A. Rept., Pt. I, 1853, p. 33; GIRARD, U. S. Explor. Exped., Herp., 1858, p. 153, pl. XIV, figs. 1–4.

Eutainia elegans, B. & G., Cat. N. A. Rept., Pt. I, Serp., 1853, p. 34 (type locality **El Dorado County, Calif.**).

Tropidonotus trivittatus, HALLOW., Proc. Ac. Nat. Sci. Phila., VI, 1853, p. 237 (**Cosumnes River, Cal.**).

Eutænia elegans ordinoides, COPE, Proc. U. S. Nat. Mus., XIV, 1891, p. 654.

Eutænia elegans brunnea, COPE, Proc. U. S. Nat. Mus., XIV, 1891, p. 654 (type locality **Fort Bidwell, Cal.**).

Eutænia elegans lineolata, COPE, Proc. U. S. Nat. Mus., XIV, 1891, p. 655 (part), (**no type**).

Eutænia infernalis infernalis, COPE, Proc. U. S. Nat. Mus., XIV., 1891, p. 657.

Eutænia infernalis vidua, COPE, Proc. U. S. Nat. Mus., XIV, 1891, p. 658 (type locality **San Francisco, Cal.**).

Tropidonotus ordinatus var. *infernalis*, BOUL., Cat. Snakes Brit. Mus., I, 1893, p. 207 (part).

Thamnophis infernalis, STEJN., N. A. Fauna, No. 7, 1893, p. 210.

Description.—Head distinct from neck, flat-topped, with rather narrow, rounded snout, and temporal regions sometimes swollen. Rostral large, bounded behind by internasal, anterior nasal, and first labial plates. Plates on top of head, a pair of internasals, a pair of prefrontals, supraocular of each side, a frontal, and a pair of parietals. Anterior and posterior nasals distinct. One loreal. One or rarely two preoculars and from two to four postoculars. Temporals normally one followed by two, sometimes 1–1 or 1–3. Eight, or very rarely seven, nine, or ten, superior and ten or eleven inferior labials, fifth or sixth superior and fifth, sixth, or seventh inferior largest, fourth and fifth or third and fourth superior reaching eye, first pair of inferior meeting on median line. Two pair of geneials, posterior equal to or little or much longer than anterior. Scales on body in nineteen or twenty-one rows. Anal plate undivided. Gastrosteges varying in number from one hundred and forty-four to one hundred and seventy-three. Urosteges in two series of from fifty-seven to eighty-nine. Eye moderate.

The head is brown or olive above, never red. The labials are yellow or olive. The chin and throat are yellow or yellowish white. The belly is yellow or olive, sometimes washed with brick red. The dorsal line is never bluish. There are four distinct types of coloration, each of which might be considered a distinct species if compared with typical specimens of the others, but all of which pass into one another almost imperceptibly when large series are examined. These types are:

(a) Similar to *T. vagrans*, the ground color being brown with three light lines, a pair of dark nuchal blotches, and numerous black or dark brown spots along the sides. This style of coloration is seen in young only, but many of the smallest specimens are unspotted.

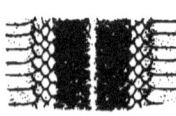 (b) Black above, with three yellowish or grayish white lines, the lateral lines sometimes blended with the color of the belly. Baird's and Girard's type of *E. elegans* was of this style.

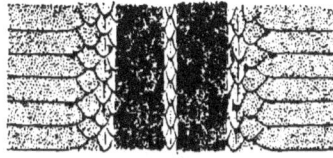

(c) Brown above, with three longitudinal lines; the dorsal line yellow; the lateral lines and more or less of the belly bright brick red.

(d) Dark olive above, with no lateral lines; the dorsal line yellow and very 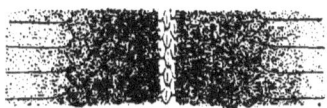 wide, the throat bright yellow, the belly deep olive or slate, with or without a yellow streak along its middle. This is Cope's *E. infernalis vidua*.

None of these forms occurs alone in any one place and the series of intergradations is complete, so that these cannot be recognized as geographical races. Nevertheless, the form (d) without lateral lines seems to occur nowhere else than on the coast slope of the San Francisco peninsula.

Length to anus..................175   450   475   510   540   560
Length of tail.................... 58   155   130   178   163

Distribution.—The Elegant Garter Snake appears to be confined to California north of the Tehachapi Mountains. I have examined specimens from Shasta (Ft. Reading), Placer (Lake Tahoe), El Dorado, Tuolumne (Tuolumne Meadows), Mariposa (Tamarack Flat, Yosemite Valley), Fresno (Fresno, Pitman Creek), Tulare, San Luis Obispo, Monterey (Pacific Grove), Santa Cruz (Soquel, Wadell Creek, Glenwood), Santa Clara (Mt. Hamilton, Santa Clara, Stephen's Creek, Palo Alto), San Mateo (Pescadero, La Honda, Searsville), San Francisco (Lake Merced, Union Square), Alameda (Calaveras Valley, Oakland), Contra Costa (Mount Diablo), Marin, Lake (Middleton), and Mendocino (Pieta, Irishes,

Sherwood's) Counties. In many parts of its range it
occurs with the *T. parietalis.*

Habits.—Little known, but similar to those of *T. parietalis.*

69.—Thamnophis vagrans (Baird & Girard). WANDER-
ING GARTER SNAKE.

Eutainia vagrans, B. & G., Cat. N. A. Rept., Pt. I, Serp., 1853, p. 35
(type locality **California**); GIRARD, U. S. Explor. Exped., Herp.,
1858, p. 154, pl. XIV, figs. 5–10.

? *Eutaenia couchii*, KENN., U. S. Pac. R. R. Surv., X, Pt. IV, 1859,
p. 10 (type locality **Pitt River, California**).

Eutænia elegans lineolata, COPE, Proc. U. S. Nat. Mus., XIV, 1891,
p. 655 (part) (**no type**).

Eutænia elegans vagrans, COPE, Proc. U. S. Nat. Mus., XIV, 1891,
p. 656.

Tropidonotus vagrans, BOUL., Cat. Snakes Brit. Mus., I, 1893,
p. 202.

Thamnophis vagrans, STEJN., N. A. Fauna, No. 7, 1893, p. 213.

Description.—Head distinct from neck, flat-topped,
with narrow, rounded snout, and temporal regions
sometimes swollen. Rostral large, bounded behind by
internasal, anterior nasal, and first labial plates. Plates
on top of head, a pair of internasals, a pair of prefrontals, a frontal, supraocular of each side, and a pair
of parietals. Anterior and posterior nasals distinct.
One loreal. One preocular. Two to four postoculars.
Temporals normally one followed by two, sometimes
1–3. Eight superior and ten or eleven inferior labials,
sixth in each series largest, fourth and fifth superior
reaching eye, first pair of inferior meeting on median
line. Two pair of geneials, posterior equal to or shorter
than anterior. Scales on body in twenty-one rows, all
keeled. Anal plate not divided. Gastrosteges varying
in number from one hundred and fifty-two to one hundred and seventy-nine. Urosteges in two series of from
fifty-three to ninety-one. Eye moderate.

The ground color above is olive or greenish yellow or brown. Along the middle of the back runs a yellow line of varying width. On the second and third rows of scales of each side is a similar yellow line. Any or all of these lines may be very indistinct or even absent. On each side of the back are two series of alternating black spots, the upper of which often encroach upon the dorsal line. These spots sometimes unite to form a zigzag band along each side, 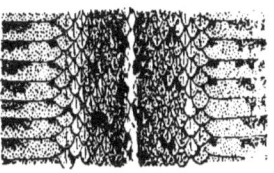 or may be obscured by the darkening of the ground color. The top of the head is usually light brown, with a yellow pineal spot. There is a pair of large dark nuchal blotches. The gastrosteges and urosteges are almost always more or less marbled with black or slate, especially near their anterior edges and along the middle of the belly. The chin and throat normally are yellowish white.

| Length to anus | 295 | 342 | 485 | 520 | 525 | 535 |
| Length of tail | 97 | 117 | 175 | 150 | 160 | 155 |

Distribution.—In California, the Wandering Garter Snake has been reported from Humboldt Bay, and is known to live on both slopes of the Sierra Nevada throughout the whole length of the chain. It ranges east across Nevada (Ash Meadows, Silver Creek, etc.) to Utah, and is common in Idaho (Salmon Mts., Birch Creek, Challis Valley, Alturas Lake, Trail Creek, Sand Point, Lewiston, Potlach Creek, Juliaetta, Wardner, Shoshone Falls, Ketcham, Blue Lake, Kootenai Co., Arco, Weiser, etc.). It crosses the border into British Columbia, and occurs in the eastern parts of Washington and Oregon, being replaced in the western portions of these states by its subspecies *T. vagrans biscutata.*

70.—Thamnophis vagrans biscutata (Cope). COPE'S GARTER SNAKE.

Eutænia biscutata, COPE, Proc. Ac. Nat. Sci. Phila., 1883, p. 21 (type locality **Klamath Lake, Oregon.***)

? *Eutænia Henshawi,* YARROW, Proc. U. S. Nat. Mus., VI, 1883, p. 152 (type locality **Ft. Walla Walla, Wash.**)

Description.—Differs from *T. vagrans* in having usually two or three (rarely one) preoculars, scales sometimes in twenty-three rows, and superior labials not infrequently seven.

The coloration usually is not different from that of typical *T. vagrans.* Some specimens, however, are so dark as to conceal the dorsal spots, and one is black everywhere excepting the chin, throat, and a few bits of skin between the scales, no lines being visible. Some dark specimens resemble, in coloration, certain examples of *T. elegans.*

Length to anus.....................375   455   500   503   550   620
Length of tail.....................132   135   155   175   175   175

Distribution.—This snake was first described from specimens collected at Klamath Lake, southeastern Oregon. I have examined specimens from that locality and from Vancouver Island, besides more than thirty from King County, Washington. Garter snakes are almost incredibly numerous in the Klamath region. That all belong to this subspecies seems improbable.

71.—Thamnophis hammondii (Kennicott). CALIFORNIA GARTER SNAKE.

Eutainia hammondii, KENN., Proc. Ac. Nat. Sci. Phila., 1860, p. 332 (type localities **San Diego, Fort Tejon, Cal.**); COPE, U. S. Explor. Surv. W. 100th Mer., V, 1875, pp. 545, 549.

Eutænia elegans couchii, COPE, Proc. U. S. Nat. Mus., XIV, 1891, p. 656.

Tropidonotus ordinatus var. *hammondii,* BOUL., Cat. Snakes Brit. Mus., I, 1893, p. 210.

Thamnophis hammondii, STEJN., N. A. Fauna, No. 7, 1893, p. 212.

---

* See Cope, Proc. U. S. Nat. Mus., xiv, 1891, p. 651.

Description.—Head distinct from neck, flat-topped, with very narrow, rounded snout and temporal regions not infrequently swollen. Rostral large, bounded behind by internasal, anterior nasal, and first labial plates. Plates on top of head, a pair of internasals, a pair of prefrontals, a frontal, supraocular of each side, and a pair of parietals. Anterior and posterior nasals distinct. One loreal. One or two or three preoculars and three postoculars. Temporals one followed by two or three. Eight (rarely seven) superior and ten inferior labials, sixth in each series largest, fourth and fifth superior reaching eye, first pair of inferior meeting on median line. Two pair of geneials, posterior much longer than anterior. Scales on body in twenty-one, or rarely nineteen, rows. Anal plate not divided. Gastrosteges varying in number from one hundred and fifty-nine to one hundred and seventy-three. Urosteges in two series of from sixty-eight to eighty-five. Eye large.

The ground color is grayish brown or olive marked, in young specimens, with numerous black spots which usually disappear with age. The dorsal line is almost always absent, or represented by a yellow spot or short line on the neck, but in some specimens extends to the tail. The lateral lines are either distinct or blend-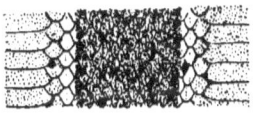ed with the color of the belly. Black spots are frequently present on the first row of scales and tips of the gastrosteges. The top of the head is olive, with a yellow pineal spot on the line between the parietal plates. Dark nuchal blotches are present. The lower surfaces are whitish or grayish yellow, sometimes barred with black between the plates, and rarely with a slight central shading of slate posteriorly.

| Length to anus | 304 | 382 | 440 | 500 | 670 | 870 |
| Length of tail | 93 | 109 | 120 | 143 | 180 | 275 |

Distribution.—Hammond's Garter Snake is confined to southern California, where it has been taken in San Diego (San Diego, Agua Caliente), Riverside (Hemet Valley, San Jacinto), San Bernardino (Ontario), Los Angeles (Los Angeles), Ventura (Santa Paula), Inyo (Owen's Valley), Kern (Fort Tejon, Kern River), Tulare (Kern River Lakes, Trout Meadows), and Fresno (Fresno) Counties.

Habits.—Like other members of its genus, this snake swims well and is usually found in or near water. Its food consists mainly of aquatic animals, such as fish, frogs, and tadpoles. One specimen was caught with a good-sized trout in its teeth. Captive individuals sometimes change their colors very quickly, in accordance with the lightness or darkness of the objects upon which they rest.

Family XIII. CROTALIDÆ.

The *Crotalidæ* or Pit Vipers are represented on the Pacific Coast and in the Great Basin by six kinds of rattlesnakes. These are our only poisonous serpents, and may be distinguished from the harmless forms by their possession of a pit in the side of the face between the eye and the nostril, and a horny, segmented rattle at the tip of the tail. They are provided with large plates along the belly, and the head is covered with small scales. The eye is well developed, with vertical pupil. There are no rudiments of limbs. Both jaws bear teeth, and near the front of the upper jaw are large, perforate, erectile poison-fangs.

Genus 37. CROTALUS.

Crotalus, LINNÆUS, Syst. Nat., 10 ed., 1758, 1, p. 214 (type horri-
dus); "*Crotalophorus*, HOUTTUYN, Linn. Nat. Hist., VI, 1764,
p. 290 (same type);" *Caudisona*, LAURENTI, Syn. Rept., 1768,
p. 92 (same type); "*Crotalinus*, RAFINESQUE, Am. Month.
Mag., III, 1818, (p. 446), IV, p. 41" (same type); "*Uropsophus*,
WAGLER, Syst. Amph., 1830, p. 176 (type triseriatus)"; *Uro-
crotalon*, FITZINGER, Syst. Rept., 1843, p. 29 (type durissus);
Aploaspis, COPE, Proc. Ac. Nat. Sci. Phila., 1861, p. 206 (type
lepida); *Æchmophrys*, COUES, Surv. W. 100th Mer. V, 1875, p.
609 (type cerastes).

The head is broad and low, with flattened top, and is
very distinct from the neck. Its upper surface is covered with scales, which are small except sometimes on
the snout. The anal plate and most or all of the urosteges are undivided. The tail is short and ends in a
horny rattle or button. The dorsal and most of the
lateral scales are keeled.

SYNOPSIS OF SPECIES.

a.—Rostral in contact with anterior nasal.
    b.—Outer edge of supraocular plate normal, not raised into a horn-like
    process.
        c.—Dark bands on tail black; light postocular line, if present, reach-
        ing row of scales next above supralabials in front of corner of
        mouth........................C. ruber.—p. 226.
        c^2.—Dark bands on tail brown (rarely in part blackish); light postocu-
        lar line, if present, reaching row of scales next above supralabials
        at corner of mouth if at all.
            d.—Rostral not wider than high.
                e.—Light postocular line one scale wide; dark postocular band
                arising below anterior corner of eye
                        C. confluentus.—p. 218.
                e^2.—Light postocular line more than one scale wide; dark postoc-
                ular band, if distinct, arising below middle of eye.*
                        C. lucifer.—p. 216;
            d^2.—Rostral wider than high..........C. tigris.—p. 220.
    b^2.—Outer edge of supraocular raised into a horn-like process.
                        C. cerastes.—p. 222.
a^2.—Rostral separated from anterior nasal by granular scales.
                        C. mitchellii.—p. 224.

---

* A narrow dark line sometimes runs forward from this point.

72.—Crotalus lucifer Baird & Girard. PACIFIC RATTLE-SNAKE.

?Crotalus oregonus, HOLBROOK, N. A. Herp., 2 ed., III, 1842, p. 21, pl. III (type locality Columbia River).

Crotalus lucifer, B. & G., Proc. Ac. Nat. Sci. Phila., VI, 1852, p. 177 (type locality Oregon and California).

Crotalus confluentus, YARROW, Surv. W. 100th Mer., V, 1875, p. 530 (part); BOULENGER, Cat. Snakes Brit. Mus., III, 1896, p. 576 (part).

Crotalus adamanteus atrox, STREETS, Bull. U. S. Nat. Mus., No. 7, 1877, p. 40.

Crotalus confluentus lucifer, COPE, Proc. Ac. Nat. Sci. Phila., 1883, pp. 11, 19, 22.

Crotalus lecontei, HALLOW., Rep. Pac. R. R. Surv., X, 1859, p. 18.

"Crotalus adamanteus var. lucifer, JAN, Elenco Sist. Ofid., 1863, p. 124."

Crotalus oregonus var. lucifer, GARMAN, Mem. Mus. Compr. Zool., VIII, 1883, p. 173.

Crotalus confluentus lecontei, COPE, Proc. U. S. Nat. Mus., XIV, p. 602.

Description.—Moderately large. Head broad, flat-topped, varying in outline according to position of fangs, etc. Rostral much higher than wide, in contact with anterior nasal. Two nasals. Usually two preoculars and four internasals. A large scale just in front of supraocular and occasionally large scales on prefrontal region. Supraocular large but not raised into a horn-like process, separated from its fellow by three to nine irregular rows of scales. Fourteen to seventeen superior and fifteen to seventeen inferior labials, first pair of latter in contact on median line in front of single pair of geneials. Two to four rows of scales between supralabials and eye. Scales in twenty-five to twenty-seven rows, keeled except in one to three rows of each side. Gastrosteges varying from one hundred and sixty-three to one hundred and eighty-nine. Urosteges sixteen to twenty-six, a few sometimes divided.

The ground color is brown, olive, gray, or dull yellow,

marked along the back with a series of large dark brown blotches which become cross-bars or incomplete rings posteriorly. These blotches are often paler centrally than about their edges and vary greatly in shade, shape, amount of separation, and contrast with the ground color. Smaller alternating blotches are usually present on the sides. Many of the scales between or around the dark dorsal blotches are light—yellow, gray, or white. These colors often show between the lateral dark blotches also. Young specimens show a light transverse streak on the supraocular, usually not present in adults. A dark streak runs from the eye to the corner of the 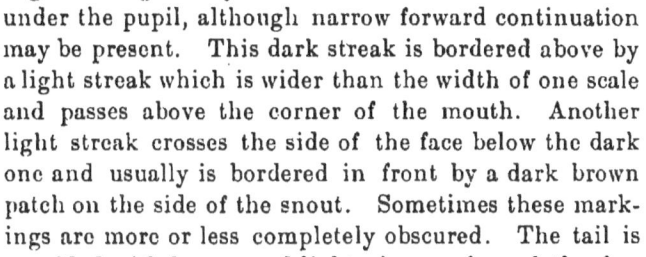 mouth, the line of its lower edge striking the eye about under the pupil, although narrow forward continuation may be present. This dark streak is bordered above by a light streak which is wider than the width of one scale and passes above the corner of the mouth. Another light streak crosses the side of the face below the dark one and usually is bordered in front by a dark brown patch on the side of the snout. Sometimes these markings are more or less completely obscured. The tail is provided with brown and light rings, a few of the former, near the tip of the tail, being occasionally blackish. The lower surfaces are white or yellow, more or less spotted or clouded with brown.

Length to anus..................253	540	740	810	810	960
Length of tail to base of rattle... 22	35	43	64	85	77

Distribution.—The Pacific or " Black " Rattlesnake occupies all parts of California except the Colorado and Mojave Deserts. It ranges across Oregon and Washington to British Columbia. Farther east, it occurs in Idaho, along the Snake River, and has been taken in many parts of Nevada and even in Utah. Throughout most of this territory it is the only rattlesnake, but in southern California is found with *C. ruber*, and possibly meets *C. confluentus* in Idaho. In California it ranges from the floor of the San Joaquin Valley up at least to an altitude of 8,600 feet in the Sierra Nevada. I have examined specimens from San Diego (De Luz, Bonsall), Riverside (San Jacinto), San Bernardino, Kern (Delano), Tulare (Kern River, Sheep Meadows), Fresno, Monterey (Monterey, Jolon), Santa Cruz (Santa Cruz, Glenwood), Santa Clara (Black Mountain, Smith Creek, Mt. Hamilton, Los Gatos, Gilroy), Sonoma (Petrified Forest), Lake (Lower Lake), Counties, California; Klamath Falls, Oregon; and Twin Falls and Blue Lakes Cañon, Idaho.

Habits.—We have no special knowledge of the habits of this species. For an account of the rattlesnakes in general see Stejneger, Report of the U. S. National Museum for 1893.

73.—Crotalus confluentus Say. PRAIRIE RATTLESNAKE.

" ?*Crotalinus viridis*, RAFINESQUE, Am. Month. Mag., IV, 1818, p. 41."

Crotalus confluentus SAY., Long's Exped. Rocky Mts., II, p. 48; BAIRD & GIRARD, Cat. N. A. Rept., Pt. I, Serp., 1853, p. 8; STEJNEGER, N. A. Fauna, No. 5, 1891, p. 111; STEJNEGER, Rep. U. S. Nat. Mus., 1893 (1895), p. 440; BOULENGER, Cat. Snakes Brit. Mus., III, 1896, p. 576 (part).

Crotalus lecontei, HALLOW., Proc. Ac. Nat. Sci. Phila., VI, 1852, p. 180 (type locality **Cross Timbers**).

?*C. lucifer* var. *cerberus*, COUES, Surv. W. 100th Merid., V, 1875, p. 607 (type locality **San Francisco Mts., Ariz.**).

Crotalus confluentus var. *pulverulentus*, COPE, Proc. Ac. Nat. Sci. Phila., 1883, p. 11 (type locality vic. **Lake Valley, New Mexico**).

Description.—Moderate. Head broad, flat-topped, varying in outline according to position of fangs, etc. Rostral much higher than wide, in contact with anterior nasal. Two nasals. Usually two preoculars and four internasals. A large scale just in front of supraocular and occasionally large scales on prefrontal region. Supraocular large but not raised into a horn-like process,* separated from its fellow by from three to six irregular rows of scales. Fourteen to seventeen superior and a similar number of inferior labials; first pair of latter meeting in front of single pair of geneials. Two to four rows of scales between supralabials and eye. Scales in twenty-five to twenty-nine rows, keeled except in one to three rows of each side. Gastrosteges varying from one hundred and seventy-eight to one hundred and eighty-eight. Urosteges nineteen to twenty-eight.

The ground color is grayish or yellowish brown, marked along the back with a series of large, darker brown blotches which usually become cross-bars or incomplete rings posteriorly. These blotches are paler centrally than about their edges and vary greatly in shape and amount of separation. Smaller alternating dark blotches are usually present on the sides. Many of the scales between or around the dark dorsal blotches are light—gray or white. These colors sometimes show between the dark lateral blotches also. Many specimens show a light transverse streak on the supraocular. A dark streak runs from the eye to the corner of 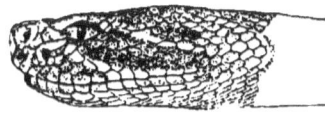 the mouth, the line of its lower edge passing in front of the eye or striking it about under its anterior corner. This dark streak is bordered above by a light streak which is not wider than the width of one scale and passes above

---

*Garman mentions a specimen with horn-like supraocular.

the corner of the mouth. Another narrow light streak crosses the side of the face below the dark one, and is bordered in front by dark brown on the side of the snout. The tail is provided with brown and light rings or cross-bars, a few of the former, near the tip of the tail, being sometimes blackish. The lower surfaces are dull yellow or white, sometimes clouded with brown.

"The color varies greatly, being sometimes duller, sometimes brighter, lighter or darker, depending upon age, season, condition of skin, climate, and the predominating color of surroundings." *

Length to anus...595
Length of tail to rattle.. 37

Distribution.—"Broadly speaking, the Prairie Rattlesnake occupies the area bounded in the East by the ninety-sixth meridian and the Upper Missouri Valley; by the main divide of the Rocky Mountains in the West; by the thirty-third parallel in Texas and the Mexican boundary further west in the South; and by the fiftieth parallel in the north. * * * Although the main divide of the Rocky Mountains in this northern region seems to be the limit of its extension to the west, yet in at least one place where there is no high crest to obstruct its passage across, it has been found on the western slope, viz.: at Lemhi, Idaho."† It has not been taken in Nevada, nor in any of the Pacific States.

74.—Crotalus tigris Kennicott. TIGER RATTLESNAKE.

Crotalus tigris, KENN., U. S. Mex. Bound. Surv., II, Rept., 1859, p. 14, pl. IV (type locality Sierra Verde and Poso Verde); STEJNEGER, N. A. Fauna No. 7, 1893, p. 214; STEJNEGER, Report U. S. Nat. Mus., 1893, 1895, p. 449; BOULENGER, Cat. Snakes, Brit. Mus. III, 1896, p. 580 (part).

---

* Stejneger. Report U. S. Nat. Mus. for 1893.
† Stejneger. Report U. S. Nat. Mus. for 1893.

Description.—Of moderate size. Head rather small, flat-topped, varying in outline according to position of fangs, etc. Rostral wider than high, in contact with anterior nasal. Two nasals. Usually two preoculars and four internasals. Supraocular large but not raised into a horn-like process ; separated from its fellow by five to seven irregular rows of scales. About fourteen superior and thirteen to fourteen inferior labials ; first pair of latter meeting in front of single pair of geneials. Three rows of scales between labials and eye. Scales in twenty-one to twenty-five rows, dorsals strongly keeled. Gastrosteges varying at least from one hundred and seventy-eight to one hundred and eighty-one. Urosteges nineteen to twenty-one.

The ground color is yellowish ash, varying from whitish to tawny, marked along the back with a series of rather small and indistinct brown blotches which become cross-bars or stripes posteriorly (whence the name *tigris*). These blotches are paler centrally than about their edges and. sometimes are nearly obsolete. Smaller alternating blotches are present on the sides. "The head markings are rather indistinct, especially the postocular stripe, which is often lost in the dense sprinkling of minute black dots covering the sides of the head." The lower surfaces are yellow or white, sometimes faintly clouded with brown.

Length to anus.. 540
Length of tail to rattle.. 35

Distribution.—The Tiger Rattlenake " was formerly only known from a few localities in southern Arizona near the Mexican boundary, until in 1891 the Death Valley exploration under Dr. Merriam extended its range very materially into the desert mountains of southern California and Nevada south of the thirty-seventh parallel, from Owen's Valley to the Great Bend of the Colorado." The vertical range is at least from 2,000 to 6,500 feet above sea-level. Some of the localities at which this snake has been taken are: Rocky Creek, Independence Creek, Owens Valley, Coso Valley, Argus Range, Panamint Mountains, and Slate Range, California, and Vegas Valley, Vegas Wash, Indian Spring Valley, and Grapevine Mountains, Nevada.

Habits.—This snake seems to be of partially nocturnal habits. It feeds upon small mammals, such as kangaroo rats and pocket mice. It probably mates in April.

75. — Crotalus cerastes Hallowell. Horned Rattlesnake. Sidewinder.

Crotalus cerastes HALLOW., Proc. Ac. Nat. Sci. Phila., 1854, p. 95 (type locality "**Borders of the Mohave River, and in the desert of the Mohave**") BAIRD, U. S. Mex. Bound. Surv., II, 1859, p. 14, pl. III; STEJNEGER, N. A. Fauna, No. 7, 1893, p. 216.

Æchmophrys cerastes, COUES, Surv. W. 100th Merid., V, 1875, p. 609.

Description.—Small. Head rather narrow, flat-topped, varying in outline according to position of fangs, etc. Rostral as broad as, or broader than, high, in contact with anterior nasal. Anterior and posterior nasals united, at least above nostril. Usually two preoculars and two internasals. Supraocular very large, raised into a horn-like process, separated from its fellow by from four to six irregular rows of scales. Eleven to thirteen superior and twelve to thirteen inferior labials, first pair 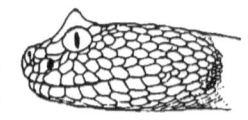 of latter in contact on median line in front of single pair of geneials. Two or three rows of scales between supralabials and eye. Scales in twenty-one rows, feebly keeled except in one to three lower rows on sides where smooth; those near middle of back with central tubercular swellings. Gastrosteges varying at least from one hundred and thirty-four to one hundred and forty-six. Urosteges sixteen to twenty-one, a few sometimes divided.

The general color above is gray, often with a yellowish or vinaceous tinge, with a series of rather small and indefinite blotches of grayish or yellowish brown. Smaller blotches or spots are usually present on the sides and on the tips of the gastrosteges. The supraocular shows an indistinct transverse streak. 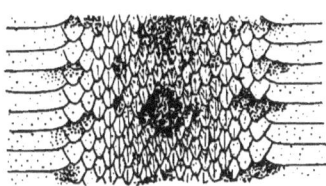 A brown streak runs from the eye to the corner of the mouth. The tail is ash-color with half rings of brown, which are much darker near its tip than anteriorly. The lower surfaces are yellowish white, sometimes faintly clouded with brown or gray.

Length to anus 415 440 450
Length of tail to rattle. 38 42 31

Distribution.—The Sidewinder occupies the lower levels of the Colorado and Mojave Deserts, where the Tiger Rattler occurs in the mountainous districts, and ranges thence across western Arizona and southern Nevada to " southwestern Utah." In California it has been taken near Salton, in the Colorado Desert; at Lone Pine, in Owen's Valley; in Panamint Valley; at Borax Flat; at Bennett Wells, Death Valley; and at Pilot Knob. It is known to occur in Nevada in Pahrump, Vegas, and Indian Spring Valleys, at Sarcobatus Flat, in the Amargosa Desert, and in the valleys of the Virgin and Lower Muddy.

Habits.—In certain parts of its range, this species is very numerous. "Its local name is derived from its peculiar mode of progression: when disturbed it moves away sideways, keeping its broadside toward the observer, instead of proceeding in the usual serpentine manner. * * * * One was shot containing a kangaroo rat (*Dipodomys*) and two pocket mice (*Perognathus*). * * * * During the latter part of April and the early part of May these rattlesnakes were often found in pairs and were doubtless mating. At such times they remained out in plain sight over night instead of retreating to holes or shelter under desert brush, and on two occasions they were found by us on cold mornings so early that they were too chilled to move until considerably disturbed." *

76—Crotalus mitchellii Cope. BLEACHED RATTLESNAKE.

Caudisona mitchellii, COPE, Proc. Ac. Nat. Sci. Phila., 1861, p. 293 (type locality **Cape St. Lucas, Lower California**).

Caudisona pyrrha, COPE, Proc. Ac. Nat. Sci. Phila., 1866, pp. 308, 310 (type locality **Canon Prieto near Ft. Whipple, Arizona**); COUES, Surv. W. 100th Merid., V, 1875, p. 608, pl. XXII.

Crotalus mitchellii, COPE, Bull. U. S. Nat. Mus., No. 1, 1875, pp. 33, 92; VAN DENBURGH, Proc. Cal. Acad. Sci., (2), IV, 1894, p. 450; *Id. ibid.,* V, 1895, p. 159; STEJNEGER, Rep. U. S. Nat. Mus., 1893 (1895), p. 454.

Crotalus pyrrhus, STEJN., W. Am. Scient., VII, April, 1891, p. 165.

Crotalus Mitchellii pyrrhus, STEJN., Rep. U. S. Nat. Mus., 1893 (1895), p. 456.

Description.—Moderately large. Head rather small, with flattened top, varying in outline according to position of fangs, etc. Rostral either higher than wide or wider than high, separated from anterior nasal by one or two rows of granular scales. Usually two nasals. Supraocular large, somewhat projecting laterally, separated from its fellow by from four to eight scales.

---

*Merriam, N. A. Fauna No. 7, 1893, p. 217.

Fourteen to seventeen superior and fourteen to eighteen inferior labials, first pair of latter meeting in front of a single pair of geneials. Three to five rows of scales between supralabials and eye. Scales in twenty-five or twenty-seven rows, keeled except sometimes in one or two rows of each side. Gastrosteges varying from one hundred and fifty-eight to one hundred and ninety-eight. Urosteges seventeen to twenty-seven, a few of the posterior sometimes divided.

The general color is white, gray, yellow, vinaceous-cinnamon, or salmon-red, minutely dotted with black or brown, and with a series of indefinite brown, black, or red blotches along the back. These dots and dorsal blotches,

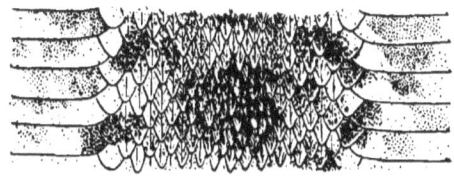

as well as smaller blotches which sometimes are present on the top of the head and on the sides, may be so faint as to cause the animal to be called the White Rattlesnake, or so dark as to produce a blackish effect; the blotches, however, never have definite outlines, appearing only as darker portions of the general "pepper and salt" style of coloration. A dark band sometimes runs down and back from the eye. The tail is gray, with black cross-bars. The lower surfaces are white or yellow, usually more or less clouded with brown.

Length to anus.. 710   770   810   840   870   930
Length of tail to rattle............. 62   74   72   60   71   90

Distribution.—This rattlesnake has been found in the Colorado Desert, near Mountain Springs, San Diego

County, California, and in the Mojave Desert. It has been taken in Arizona, and ranges the whole length of the peninsula of Lower California.

Habits.—This seems to be distinctively a desert species. Like other rattlesnakes, it is ovoviviparous. A specimen taken at San José del Cabo, in September, contained three young about 260 *mm.* in length.

77.—Crotalus ruber (Cope). WESTERN DIAMOND RATTLE-SNAKE.

Crotalus adamanteus ruber, COPE, Proc. U. S. Nat. Mus., XIV, 1891, p. 690 (type locality unknown.)

Crotalus atrox ruber, STEJNEGER, Rep. U. S. Nat. Mus., 1893 (1895), p. 439.

Crotalus ruber, VAN DENBURGH, Proc. Cal. Acad. Sci. (2), V, 1895, p. 1007.

Description.—Large. Head broad, flat-topped, varying in outline according to position of fangs, etc. Rostral usually higher than wide, in contact with anterior nasal. Two nasals. Usually two preoculars and two to four internasals. A large scale just in front of supraocular. Supraocular large but not raised into a horn-like process; separated from its fellow by six or seven irregular rows of scales. About sixteen or seventeen superior and seventeen to nineteen inferior labials; first pair of latter not meeting on median line in front of single pair of genials. About four rows of scales between supralabials and eye. Scales in twenty-seven to twenty-nine rows, of which one or two on each side are smooth. Gastrosteges varying from one hundred and eighty-six to one hundred and ninety-nine. Urosteges twenty-two to twenty-six.

The general color is light red, reddish cinnamon, or brownish yellow, with a series of large, darker blotches along the back. These blotches are sometimes very

indefinite, especially toward the sides.   On the middle of
the back they are separated by light yellow or white.
This light edging may or may not be continued onto
the sides, where smaller indefinite dark blotches may
often be seen.   The head is unicolor above.   A faint

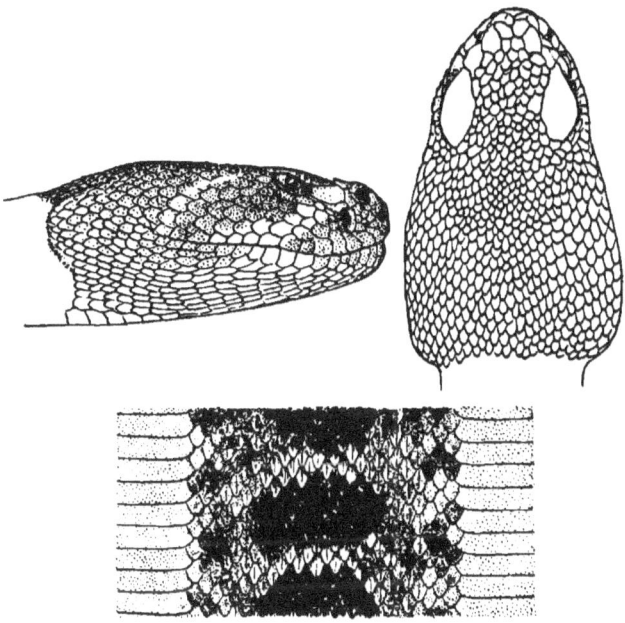

light stripe crosses the side of the face from the preocular plates to the mouth.   The scales behind and
above this light stripe are a little darker than the ground
color and sometimes are set off posteriorly by a light line
running down and back from the posterior corner of the
eye and striking the supralabials in front of the corner
of the mouth.   The tail is ash-color, with from three to

five black rings or cross-bars. The lower surfaces are
yellowish, often faintly clouded with light brown.

Length to anus.. 935   1080
Length of tail to rattle..................... 55   75

Distribution.—This, the largest rattlesnake of the
West, was first described from a specimen of unknown
origin. It has since been found on the western slopes
of San Diego and Riverside Counties, at Twin Oaks,
San Jacinto, and De Luz.

INDEX.